Spatial Demand Allocation for Distribution System Design

The mission of the Awwa Research Foundation (AwwaRF) is to advance the science of water to improve the quality of life. Funded primarily through annual subscription payments from over 1,000 utilities, consulting firms, and manufacturers in North America and abroad, AwwaRF sponsors research on all aspects of drinking water, including supply and resources, treatment, monitoring and analysis, distribution, management, and health effects.

From its headquarters in Denver, Colorado, the AwwaRF staff directs and supports the efforts of over 500 volunteers, who are the heart of the research program. These volunteers, serving on various boards and committees, use their expertise to select and monitor research studies to benefit the entire drinking water community.

Research findings are disseminated through a number of technology transfer activities, including research reports, conferences, videotape summaries, and periodicals.

Spatial Demand Allocation for Distribution System Design

Prepared by:
Thomas W. Chesnutt, Ph.D.
A&N Technical Services, Inc., Encinitas, CA 92024
and
Los Angeles Department of Water and Power
With
Metropolitan Water District of Southern California
Irvine Ranch Water District
Calleguas Municipal Water District

Jointly sponsored by:
Awwa Research Foundation
6666 W. Quincy Avenue
Denver, CO 80235-3098
and
LADWP
PO Box 51111
Los Angeles, CA 90051-0100

Published by the

Awwa Research Foundation

Advancing the Science of Water®

IWA Publishing

DISCLAIMER

This study was funded by the Awwa Research Foundation (AwwaRF). AwwaRF assumes no responsibility for the content of the research study reported in this publication or for the opinions or statements of fact expressed in the report. The mention of trade names for commercial products does not represent or imply the approval or endorsement of AwwaRF. This report is presented solely for informational purposes.

Published by IWA Publishing, Alliance House, 12 Caxton Street, London SW1H 0QS, UK

Telephone: +44 (0) 20 7654 5500; Fax: +44 (0) 20 7654 5555; Email: publications@iwap.co.uk;
Web: **www.iwapublishing.com**

AwwaRF report number 90934F. Originally published by AwwaRF for its subscribers in 2003.
IWA Publishing version published 2004.

ISBN 1 84339 846X

CONTENTS

TABLES

FIGURES

FOREWORD

The Awwa Research Foundation is a nonprofit corporation that is dedicated to the implementation of a research effort to help utilities respond to regulatory requirements and traditional high-priority concerns of the industry. The research agenda is developed through a process of consultation with subscribers and drinking water professionals. Under the umbrella of a Strategic Research Plan, the Research Advisory Council prioritize the suggested projects based upon current and future needs, applicability, and past work; the recommendations are forwarded to the Board of Trustees for final selection. The foundation also sponsors research projects through the unsolicited proposal process; the Collaborative Research, Research Applications, and Tailored Collaboration programs; and various joint research efforts with organizations such as the U.S. Environmental Protection Agency, the U.S. Bureau of Reclamation, and the Association of California Water Agencies.

This publication is a result of one of these sponsored studies, and it is hoped that its findings will be applied in communities throughout the world. The following report serves not only as a means of communicating the results of the water industry's centralized research program but also as a too to enlist further support of the nonmember utilities and individuals.

Projects are managed closely from their inception to the final report by the foundation's staff and large cadre of volunteers who willingly contribute their time and expertise. The foundation serves a planning and management function and awards contracts to other institutions, such as water utilities, universities, and engineering firms. The funding for this research effort comes primarily from the Subscription Program, through which water utilities subscribe to the research program and make an annual payment proportionate to the volume of water they deliver and consultants and manufacturers subscribe based on their annual billings. The program offers a cost-effective and fair method for funding research in the public interest.

A broad spectrum of water supply issues is addressed by the foundation's research agenda; resources, treatment and operations, distribution and storage, water quality and analysis, toxicology, economies, and management. The ultimate purpose of the coordinated effort is to assist water suppliers to provide the highest possible quality of water economically and reliably. The true benefits are realized when the results are implemented at the utility level. The foundation's trustees are pleased to offer this publication as a contribution toward that end.

Many water utilities are facing a period of growth as populations increase and communities expand. In order to meet increasing water demands, it is imperative that water utilities use state of the art technologies to design and plan new or rehabilitated water distribution systems. New technologies of automated metering help utilities optimize distribution system networks and model water quality in the distribution system. The Foundation undertook this research project in order to address the issues to the spatial demand allocation of water demands used for distribution system planning and design.

Edmund G. Archuleta, P.E.
Chair, Board of Trustees
Awwa Research Foundation

James F. Manwaring, P.E.
Executive Director
Awwa Research Foundation

ACKNOWLEDGMENTS

The authors of this reported express their gratitude to the water utilities that contributed expertise, experience, and data to this research effort:

Los Angeles Department of Water and Power – Paul Liu, Terri Koch, Greggory Ammon, Steve Ho, Paul Gillis, and Richard West
Metropolitan Water District of Southern California – Timothy Blair and Warren Teitz
Irvine Ranch Water District – Gregory Heiertz
Calleguas Municipal Water District – Donald R. Kendall

The authors also gratefully acknowledge the members of the Project Advisory Committee: Nick Braybrooke (Las Vegas Valley Water District), Lee Cesario (Denver Water), and Randy Hill (Victor Valley Water District). Warren Teitz contributed drafts of Chapter 3, numerous other insights, and considerable patience. Finally, the authors thank Sanjay Gaur for research assistance related to this project. The authors especially appreciate the advice and guidance of AwwaRF project manager, Ms. Traci Case.

EXECUTIVE SUMMARY

With the advent of new information technologies—geographical information systems, automated meter reading, and organizationally integrated data—water system managers and planners have new opportunities to analytically confront distribution system design and planning issues. Information technology has been changing at a rapid pace. New methods of automated metering will open new possibilities for network optimization of new or rehabilitated distribution systems and flow modeling of water quality.

This research project addresses the issues of the spatial allocation of water demands used for distribution system planning and design. Water system demand characteristics (quantity, quality, timing, and location) drive many distribution system management, operation, and design decisions. Most notably, characteristics such as average-annual use, maximum-day use, maximum-hour use, and fire-flow requirements constrain the makeup, size, and timing of distribution system modifications and expansions. Though very sophisticated network models have been developed to design and analyze distribution system modifications and expansions, these models traditionally treat nodal demand characteristics as a given, fixed and known. However, in many cases, spatial distribution of demand is not well defined. Given its importance in water system management, operation, and design, the overall goal of this research is to quantify and further understand the spatial characteristics of water demand.

RESEARCH OBJECTIVES

1. Assess existing tools and methodologies used for identification, prediction, and spatial allocation of water demands for distribution system planning and design.
2. Make use of state-of-the-art technologies, water consumption data, and other regional data to develop improved empirical depictions of the spatial distribution of water demand.
3. Develop common tools and methodologies for predicting and allocating demand in water distribution planning and design models.
4. Describe methods for quantifying uncertainties associated with water demand predictions used for system planning and design.

APPROACH

1. Research existing tools and methodologies for quantifying spatial distribution of water demand and allocating demand in distribution system planning models.
2. Explain why data and sampling issues are relevant to spatial demand allocation, provide steps for sample design, and illustrate using utility data.
3. Using appropriate statistical methods, specify and estimate models for water consumption data that can be used to depict the empirical shape of water consumption in time and space. Use these depictions for developing modeling tools and methodologies for predicting and allocating water demands and quantifying water use prediction uncertainties.

4. Test and refine spatial demand analysis methods and tools on participant system and demand data.
5. Publish methods for use in system planning and design.

FINDINGS

1. Water consumption data, though messy, can provide spatially more precise estimates of demand load for use in distribution system models.
2. Attention to data quality and sample design permit improved "inference"—that is, conclusions can be correctly drawn.
3. Statistical methods and models provide a useful means to
    a. Interpolate average daily consumption through time
    b. Provide a formal basis for probabilistic statements about demand load
    c. Formally test for the existence of spatial differences and trends in consumption
    d. Quantify the effect of weather on demand load (maximal effect within consumption sample and maximal effect throughout recorded weather history.)
4. Additional applications of these methods include:
    a. Incorporating long-term water demand growth drivers into distribution system planning.
    b. Using consumption models to populate demand load scenarios for use in more demanding network optimization models.
    c. Applying consumption models for improved quantity and quality reliability planning for water distribution systems

CHAPTER 1
SPATIAL ALLOCATION OF WATER DEMAND

PROBLEM STATEMENT

Water system demand characteristics (quantity, quality, timing, and location) drive many distribution system management, operation, and design decisions. Most notably, characteristics such as average-annual use, maximum-day use, maximum-hour use, and fire-flow requirements constrain the makeup, size, and timing of distribution system modifications and expansions. Though very sophisticated network models have been developed to design and analyze distribution system modifications and expansions, these models traditionally treat nodal demand characteristics as a given, fixed and known. However, in many cases, the spatial distribution of demand is not well defined. Given its importance in water system management, operation, and design, additional research is required to further understand the spatial characteristics of water demand. To address this need, the following project objectives were set:

1. Assess existing tools and methodologies used for identification, prediction, and spatial allocation of water demands for distribution system planning and design.
2. Make use of state-of-the-art technologies, water consumption data, and other regional data to develop improved empirical depictions of the spatial distribution of water demand.
3. Develop common tools and methodologies for predicting and allocating demand in water distribution planning and design models.
4. Describe methods for quantifying uncertainties associated with water demand predictions used for system planning and design.

BACKGROUND

Most water demand models focus on long-term aggregate water demand. These long-term models tend to be specified at an aggregate level because the data are better. While appropriate for having the highest likelihood of sorting out trends in many long-term determinates (population, employment, income, and water rates), aggregate long-term demand models are very uninformative to the problems associated with the timing and location of incremental system capacity expansions. Furthermore, studies associated with the temporal distribution of demand (diurnal residential, commercial, and industrial end uses) usually do not address related spatial allocations.

Advances in land use planning data and in the information technology required to manipulate geographic data have created the opportunity to develop closer ties between distribution system design and the spatial attributes of water demand. Clearly, the water requirements of any geographic sub-area depend on the land uses permitted in that region. This project addresses the need to develop improved empirical depictions of the spatial distribution of existing and future water demand. Additionally, modeling approaches necessary to couple geographic information system (GIS) based land use data with widely available billing histories of water consumption will be developed. In so doing, improved spatial demand predictions will be realized by connecting water demand design criteria more closely to current and planned land use data.

In rapidly developing regions, design issues related to demand characteristics continue to prevail in coming years (i.e., What kind of water main should be installed today to meet future expected growth?). However, even in relatively mature areas close to their ultimate build-out levels, significant distribution system replacement and rehabilitation issues will continually be confronted. In an increasingly competitive environment, water agencies must be attentive to ways of avoiding capital costs and to means for realizing new efficiencies. By coupling advances in information technologies with robust modeling methods, this project provides better input to distribution system planning and design, and in this way, be a critical component in assisting water agencies in delivering high levels of service at a reduced cost.

APPROACH

The technical approach to this research is as follows:

1. Research existing tools and methodologies for quantifying spatial distribution of water demand and allocating demand in distribution system planning models.
2. Explain why data and sampling issues are relevant to spatial demand allocation, provide steps for sample design, and illustrate using utility data.
3. Using appropriate statistical methods, specify and estimate models for water consumption data that can be used to depict the empirical shape of water consumption in time and space. Use these depictions for developing modeling tools and methodologies for predicting and allocating water demands and quantifying water use prediction uncertainties.
4. Test and refine spatial demand analysis methods and tools on participant system and demand data.
5. Publish study and tools for use in system planning and design.

ORGANIZATION OF THE REPORT

This report is organized to provide water utility managers and analysts a better understanding of the data quality and methodological issues surrounding the use of water consumption in water distribution system modeling. The report includes the following:

- Chapter 2 provides a literature search, insights from a small utility survey, and a review of current practices.
- Chapter 3 lays out the planning concepts of spatial demand allocation and water demand models. It specifies the areas that existing water consumption measures can add resolution to the spatial distribution of water demand.
- Chapter 4 examines issues of scientific sampling and data validation. It provides steps for sample design and illustrates with a sample design for water consumption in the Los Angeles Department of Water and Power service area.

- Chapter 5 specifies the general functional form for the statistical model of water consumption. This specification, in turn, provides the basis for formal scientific inference.
- Chapter 6 provides the estimation results for the consumption models
- Chapter 7 reviews additional applications of the formal models to planning for distribution system growth, improvements to production pattern-based interpolations of spatial demand, and planning for reliable delivery of potable water quantity and quality.
- Chapter 8 concludes the report.

Appendix A provides details of the utility survey, Appendix B provides estimated statistical models by production subareas and references and a list of abbreviations terminate the report. We turn in Chapter 2 to the literature search and review of existing practices.

CHAPTER 2
LITERATURE REVIEW AND SURVEY OF EXISTING PRACTICES

> "One of the most difficult tasks in successful hydraulic modeling of water distribution systems is the allocation of accurate representative water consumption data for the network model."
>
> –From Altman, et al. 1997

LITERATURE REVIEW

Demand allocation constitutes an important part of the hydraulic modeling process. It is one of the main steps as documented in AWWA M32 "Distribution Network Analysis for Water Utilities" published in 1989. The study mentions the use of meter-read books as a base for nodal demands within a network. Customer meters are usually read bi-monthly or quarterly so an average annual use value can be obtained for each customer. When appropriate peaking factors are applied along with diurnal curves, a model will be ready for calibration against field data. More recently, AWWA sponsored a research study by Lee Cesario titled "Modeling, Analysis, and Design of Water Distribution Systems" in which the importance of demand allocation was emphasized. One section of the book mentions that in order for a model to be simulated properly, water demand must be accurately distributed geographically and over time. These studies along with others (Lowry and Porter 1994, Orr, et. al 1994, Purves and Cesario 1993) have emphasized the importance of accurately distributing nodal demands within a network. Currently, there is not a comprehensive documented method to incorporate current information technology with billing data to empirically determine demand allocation.

Several other important studies have focused on residential water use. These studies were useful to the water industry from a consumption and time-pattern standpoint but not on spatial allocations of water demands. AwwaRF has sponsored two studies within this decade. The 1993 study (Bowen, et. al) is titled "Residential Water Use Patterns". Five cities were studied where electronic meters were installed at various residencies to determine a 24-hour demand pattern during the summer and winter months. Their main conclusions focused on the amount of water used by an individual residence and when the 24-hour peak demands occurred. It was a useful study where these patterns could be incorporated into a distribution model for an extended-period simulation.

The second study AwwaRF sponsored study was title "North American Residential End Use Study." The main focus of the progress report was to determine how water is used by residential customers, that is, where are the end uses of water. The resulting database contained over 1.5 million events and allowed a high-resolution look at water's end uses within the single family residences examined by the study. Though no explicit treatment was given to the spatial attributes of these data, additional analyses of these data could prove fruitful. The seasonal distribution of indoor to outdoor end-uses, for example, is directly relevant to defining demand for distribution system models.

Various other studies have been conducted in the past (Kuranz 1942, Graeser 1958, Hudson 1964, Sisco 1967, Neilsen 1969, Brittain 1974, Newman and Noss 1982, Yanov and Koch 1987). The main focus of these articles was to determine residential use patterns as depicted by diurnal curve patterns, actual usage within a structure, or comparing usage patterns from different parts of the country. These studies can be useful in hydraulic modeling but differ in scope from this research. None of these studies focused on developing a scientifically defensible methodology to incorporate water consumption from customer billing information using current information technology. This framework is designed to help utilities develop a database interface to accurately allocate demands within a hydraulic model. The formal depiction of demand factors from this study gives planners a more accurate way of developing facilities for new communities or revise existing demands if the land use for an area has changed. This methodology if utilized, will serve as a long-term base for water professionals to properly model their systems during various regular and emergency scenarios.

UTILITY SURVEY OF EXISTING PRACTICES

The survey of existing practices in water utilities had a very low response rate. This yielded a small potentially biased sample (the sample size of usable responses was 15 surveys) that limits what the conclusions that may be drawn to the population of all water utilities. Nonetheless, some insight may be gained by examining the responses. Additional detail on the survey can be found in Appendix A. Findings include:

- **Processes followed by utilities in generating spatial demand forecasts**
  The demand forecasting processes reported vary from:
  - rudimentary methods such as assumed fixed annual percent increase, to
  - intermediate simplicity using demographics or moderate system disaggregation,
  - to more sophisticated techniques utilizing detailed geographic sub-areas, demographics, land use, and/or water-use category breakdowns
- **Specific computer applications, modeling or statistical analyses used by utilities**
  - current statistical regressions,
  - current hydraulic modeling, and
  - movement toward more GIS and hydraulic model use
- **Organizational unit within utilities responsible for spatial demand forecasts, and/or consultants utilized by utilities to conduct spatial demand forecasting**
  - in decreasing order, units answering the survey were engineering, followed closely by planning, then operations or unknown.
  - About half the utilities use consultants in place of or in addition to staff.
- **Demand forecasting methods used by utilities**
  - In decreasing order, "Per Capita" and "Land-Use Based" are the most predominantly used methods.
- **Demand forecasting period(s) used by utilities**
  - The three most common periods are 5, 10, and 20 years.
- **Computer models/packages utilities use for hydraulic modeling**
  - The two most frequently used models are H2Onet and Cybernet/WaterCAD (WaterCAD is a later-generation evolution of Cybernet, from the same company)
- **Computer models/packages utilities use for spatial demand forecasting**
  - In decreasing order, the most commonly used software are GIS (mostly ESRI products), AutoCAD and Excel.

- **Methods used by utilities to input spatial demand forecasts into hydraulic models**
    - Manual input dominates, with import/export files closely following.
- **Variables taken into account in demand forecasts by utilities**
    - "Population Growth" is by far the most used, with much less usage of "Regional Economic Forces" and "Climate".

REVIEW OF EXISTING PRACTICES

Network models have long been used in the modeling of distribution systems. The application of linear programming methods to network flow problems occurred in the late fifties and early sixties[1]. Beyond the task of designing the minimum cost network to meet customer demand, water quality modeling issues have also driven the use of advanced network modeling methods[2]. Designing distribution systems clearly requires an accurate definition of where system demands occur in time *and* space.

Planning Constructs and Demand

To understand potential uses of consumption data in distribution system modeling, several planning constructs should be defined. Figure 2.1 depicts three planning constructs: different views of demand that vary by time. Baseline demand is the average annual demand level. Seasonal demand shows how average daily demand changes through the year. Demand is lower in the winter and increases in the summer as outdoor end uses increase. Diurnal demand is the variation of demand within a day. This tends to be low during the night and increasing during daylight hours.

---

[1] See, for example, Ford, L. R. and D. R. Fulkerson, 1962, *Flows in Networks*, Princeton University Press.

[2] The book by Clark and Grayman—*Modeling Water Quality in Drinking Water Distribution Systems*, 1998, AWWA—provides an excellent overview of the subject.

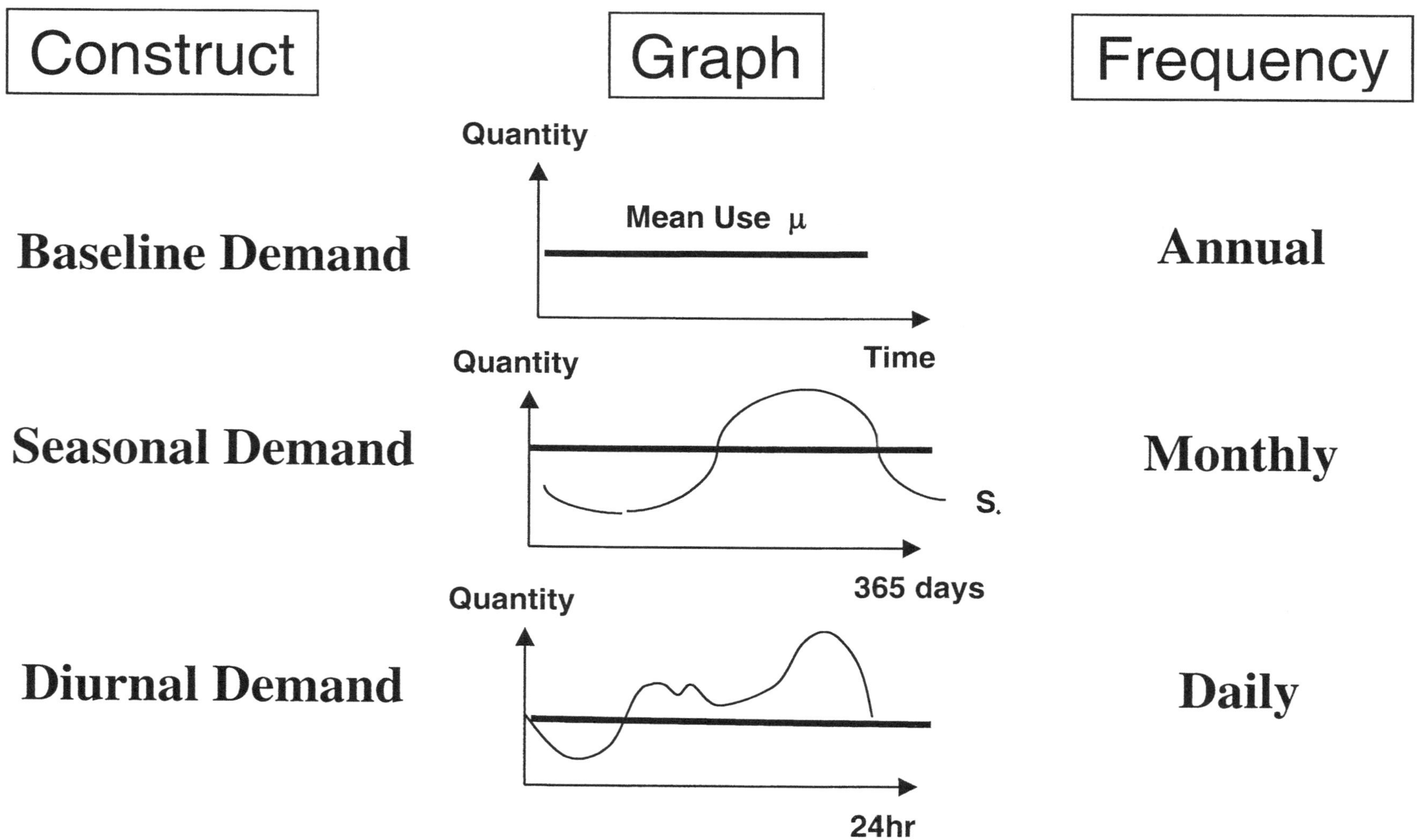

Figure 2.1: Three Planning Constructs – Baseline, Seasonal, and Diurnal Demand

Planning Constructs and Demand Data

There are two types of demand measures that can be brought to bear on distribution system issues—Production data and consumption data. Production data is a direct measure of the input into a distribution system. It is usually measured as a flow rate. Consumption data is a direct measure of the output of a distribution system. It is usually billed, and is hence measured volumetrically. As depicted in Figure 2.2, production data is characterized by

- High temporal resolution (ex. flow in cubic feet per second) and
- Poor spatial resolution (pressure zones).

Consumption data, on the other hand, is characterized by

- Better spatial resolution (connection level) and
- Poorer temporal resolution (monthly, bimonthly, and quarterly).

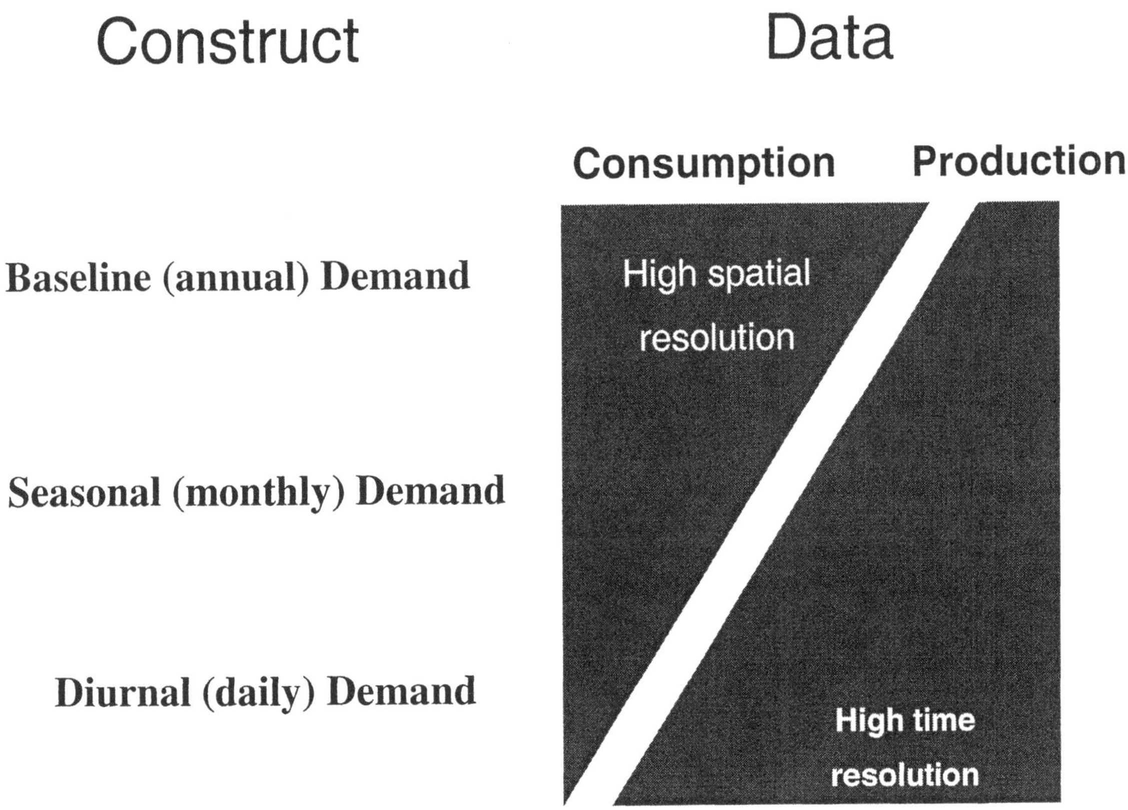

Figure 2.2 Planning Constructs and Planning Data

Planning Constructs and the Research Approach

The approach of this report focuses on the planning constructs that consumption data can most productively define. Thus, readers should expect no revealing insights on the definition of fire flow requirements. Similarly, the definition of diurnal demand is beyond reach of monthly billing system data. The picture depicted in Figure 2.2 is not set in concrete. As AMR technology becomes more widespread, consumption measures will attain a higher time resolution. The development of more precise measures of consumption should enable even more sophisticated applications of customer-driven distribution system design.

The statistical models developed within this report are designed to address two deficiencies of consumption records derived from a billing system. The first deficiency is the relatively high level of "noise" in these historical records—including meter mismeasured, meter reader misread, or other forms of data corruption. The models developed in this report can be used to identify records with data quality issues. The second deficiency is the poor time resolution of consumption records. The models used in this report develop a sophisticated depiction of seasonal demand that is permitted by the rolling nature of meter read data. These two points will be further addressed in Chapter 5.

CHAPTER 3
SPATIAL ALLOCATION AND WATER DEMAND MODELS

"Meter read book information is quite messy to work with..."
- From *Determination of Water Consumption,* Internal Los Angeles Department of Water and Power Lecture, W.E. Kruse, 1974

Chapter 2 discussed the use of demand factors and forecasts in system planning, such as baseline, seasonal, and peak demand. This chapter focuses on the formal models that have been developed to explain and forecasts water demand. We examine the different type of models and highlight those most applicable to spatial demand allocation.

This section will review the level of spatial allocation and other factors affecting model selection for water demand forecasting. Critical issues include the timeframe of the forecast, water use data availability, and supplementary data sources. Fortunately, with the rapid expansion of water related GIS systems and related data in the water industry, more geographically refined water use and related information is available for modeling than ever before. Even so, certain characteristics of water use data and model structures may limit the spatial disaggregation of water demand models.

The customer-specific consumption records within a utility's billing system are often the best data available for models of water demand. Thus, this section ends with a discussion of the problems with using billing data to derive spatially allocated demand models, as well as how some utilities, such as the Irvine Ranch Water District and Los Angeles Department of Water and Power have used GIS technology to solve these limitations.

WHY MODEL DEMAND?

Water demand models serve many purposes within water utilities. From a system planning perspective, demand models can *explain* how water use reacts to factors such as weather, price changes, conservation, income growth and other factors such as demographic change. The explanation of how these factors determine water demand can be used to *predict* how water demand can be expected to change. Formal water demand models can also provide a framework for analyzing the forecast uncertainty, helping planners make rational decisions and manage risk.

The benefits of formal water demand models can be applied directly to spatially allocated demand forecasts. Spatially allocated demand forecasts can be used in conjunction with system models to pinpoint potential system bottlenecks, increasing reliability and potentially reducing overall capacity expansion costs. Depending on the specification, demand models can help explain spatial differences in water use patterns, and exploit that information to generate more accurate baseline and peaking forecasts. This ability may be lost with "top-down" spatial demand allocation approaches, and can lead to faulty assumptions regarding the impacts of geographically varying demand growth. Finally, spatially allocated demand models provide an opportunity to perform several levels of sensitivity analysis on growth and changing water use patterns.

Types of Demand Models

Even a cursory review of water demand forecasting literature shows that utilities use a wide variety of models for water demand forecasting, ranging from simple percent growth methods to sophisticated econometric models[1]. Times series analysis is appropriate for many forecasting applications, and researchers are beginning to explore advanced mathematical techniques such as genetic algorithms, fuzzy programming and artificial neural networks to forecast water demand. GIS-based forecasting systems are becoming increasingly popular, and provide a powerful tool for spatially disaggregated demand forecasting.

The specification of a demand model for spatially allocated water forecasting is dependent on a number of factors, including:

1. The purpose of the forecast
2. The level of geographic disaggregation
3. The forecast timeframe
4. The availability of historical water use and/or billing data
5. The availability of demographic and/or land-use data
6. The availability of other GIS related data such as system information
7. The types of uncertainty to be analyzed

Purpose of the Forecast

The first question that must be asked is; what is the purpose of the forecast? The answer will suggest a model type and spatial disaggregation. A forecast designed for predicting short-term peaking might look an hour to several weeks in the future, and may include a high degree of spatial disaggregation, while a forecast used for resource planning might span 20 years and feature a low degree of spatial disaggregation. Forecast designed for developing spatially varying water rates may forecast one year out and feature a completely different level of geography. Each of these forecasts suggests a different type of model and level of spatial disaggregation. Alternatively, the data available at the desired level of spatial disaggregation may dictate model form and uses.

Short-Term Demand Forecasting

Short-term demand models produce forecasts anywhere from one hour ahead to several years in the future, and are typically used for short-term system operations and optimization, but are also used for rate setting, peak forecasting, and reservoir operations.

The spatial allocation used in short-term models varies according to purpose, ranging from pressure zones, and treatment areas to other utility-specified areas. Many short-term forecasting applications employ time-series or trending models such as exponential smoothing, which can be applied with minimal data needs to a large number of demand nodes. These

---

[1] Standard water demand references include Billings and Jones (1996), *Forecasting Urban Water Demand*, AWWA, and Baumann, Boland, and Hanemann (1988) Urban Water Demand Management and Planning.

forecasting methods, which are described extensively in the literature, are not the focus of this study and therefore are not described here.

Long-Term Demand Forecasting

Long-term water demand forecasts are generally used for long-range system planning, resource development and rate planning, and range anywhere from several years to 20 years in the future. Some land-use based water demand forecasts will project system demands to a "build-out" state, representing the theoretical limits of development based on a general plan.

In functional form, long-term forecasting models range from relatively simple per capita models to data-intensive cross-sectional times-series econometric models. Most long-term models are based on "driving" variables, such as population, households and land-use; and "explanatory" variables, which develop associated water use factors. In these model formulations, long-term projections of the driving variables (population or acreage for instance) typically obtained from an outside agency or source, are combined with annual or seasonal water use factors. The water use factors may be fixed over time (gallons per acre per day) or vary based on projected changes in explanatory variables such as weather, household demographics, price, and conservation. The properties of some of the most common long-term models discussed in the literature are summarized in Table 3.1.

Table 3.1
Summary of Long-Term Water Demand Models

Model Type	Driver Variable	Explanatory Variables	Typical Model Form	Spatial Allocation
Per-Capita	population	None	fixed water use per person	service area to custom zones
		Weather, price, income	Linear regression	
Disaggregated Factor Forecast	employment group or customer class	None	fixed factors	May be limited by billing or use data
Econometric Models	housing type, employment group or customer class	Weather, socioeconomic, price, income, conservation	Econometric models project water use factors	May be limited by billing or use data
Land-Use (duty-factor) models	land use type	None (typically)	water use per acre of land-use	service area to land-use zone

SPATIAL ALLOCATION AND WATER USE DATA

The level of geographic disaggregation of the forecast models will be determined to a great extent by the availability of spatially allocated water use data. For most utilities the options for geographically allocating water use are either system meters measuring water use in distinct zones, or geo-referenced (or system-referenced) customer billing data.

There are a number of issues arising from the use of billing data and demand forecasting, and these are summarized at the end of this chapter. Advantages to using billing data include general availability and almost unlimited geographic disaggregation. Important limitations include the monthly and bimonthly time intervals of most billing data, and problems associated with developing water use time series. It is assumed that most utilities building spatially allocated demand models will develop demand zones based on existing system meters or pumping zones. The benefits of system-based zones include no restrictions on the model timeframe, and superior peak modeling. The major limitation may be in the inability to model at a desired geographic resolution. However, this limitation may be outweighed by the ability to develop superior models.

The key question after the spatially allocated water demand zones are selected is what data is available for demand modeling. It is at this point that GIS and their spatial data analysis capabilities are crucial for developing associated data.

Spatial Allocation and GIS Systems

It is unlikely that demand zones created from system meters adhere to political or demographic boundaries such as zip codes, census tracts or specific land uses[2]. However, if the demand zones are converted into GIS coverages, data from a variety of sources can be developed using GIS analysis into all but the smallest allocation zones. The following table contains a partial list of the relevant variables that can be transformed from their native geographies – zip codes, census tracts or blocks, weather stations – into the custom boundaries used in a spatially allocated demand model:

[2] The exception might be dedicated agriculture or landscape meters.

Table 3.2
Variables Available with GIS Processing

Demographic
Population
Housing type
Housing density
Household size
Housing stock age
Water end-use data
Economic
Employment
Industry Type
Large Landscape
Income
Other
Land-use Mix
Meter Counts
Water Rate Zones
Climate
Parcel data
Conservation Activity
Saturation survey

Model Formulation and Estimation

Combining demand zone water use data with matching GIS-derived demographic, economic, weather and other data opens up the possibility for a near infinite number of possible model formulations. One possibility is developing water use factors for different types of land use, and using this information in conjunction with forecasts of land use to project disaggregated long-term demand growth. Alternatively, a cross-sectional land use model could be built regressing zonal water use against different mixes of land use combined with economic and other zonal data.

Alternatively, various types of econometric models can be constructed from the zonal water use and socioeconomic data developed in GIS system. One sophisticated example could mimic the structure of MWD-MAIN, developing econometric models for different customer classes and using demographic and economic projections, disaggregated by zone, to develop the spatially allocated demand forecasts. The advantage of such a model is the ability to capture changes water use tied to long-term and regional and local trends. The disadvantage is the sophisticated statistical modeling and complex data handling required in constructing large econometric models.

Spatial Allocation Models: Formulation and Application

The optimal form for spatially allocated models depends on the kinds of uses discussed above. How to choose the right form is discussed further in following chapters. The premise of this chapter is that system meters are available for generating satisfactory spatially allocated water demand zones. In some cases, this may not be the case. Therefore, before moving on, it is useful to review some of the issues with using billing data to construct water demand zones.

Conclusion: Spatial Demand Allocation and Billing Data

Spatially allocating customer water use or billing data into small geographic areas such as system demand nodes is now readily accomplished using GIS-based methods. However, using billing data to develop geographically-based demand nodes may be complicated by a number of factors including billing cycle properties and water use factor development. Even if only annual water use factors are needed, billing records require significant processing before valid water use models and thus demand patterns and peaking factors can be developed.

Developing spatially allocated water demand models for small geographic areas requires understanding how water use at small geographic levels changes over time. Is the water use at certain geographies more or less able to be explained by traditional water demand models? On one hand, spatially allocated water demand models may be robust if the geographic areas represent stable neighborhoods with homogenous water use patterns or encompass a large enough area to develop water use models from land-use or demographic variables. On the other hand, spatially allocated demand models at very small geographies may develop unstable demand patterns or statistically insignificant water demand models. In other words, there may be a geographic disaggregation limit where the problems inherent in developing water use models using small areas outweighs the benefits of the fine spatial allocation.

Spatial Allocation and Billing Cycles

Water utilities typically bill customers on either a monthly or bi-monthly basis, and may also employ rolling billing cycles, where different sets of customer groups are billed on different days of the month. Thus, unless only annual water demand factors are to be developed, developing meaningful monthly and peak factors and forecasts at small geographies requires understanding how and when the meters are read and water use is reported. A crucial issue in building spatially allocated demand models is whether the small geographies aggregate customers within the same billing cycles.

This is because water use data derived from rolling meter reading or billing cycles may include water deliveries from one month prior if billed monthly and two months prior in bi-monthly billing systems. Figure 3.1 shows a hypothetical example of a monthly billing system with two billing cycles: one mid month, and one end-of-month. In this example, the total billed water use in April will include water use from March from the first cycle, while some April water use is actually billed in May.

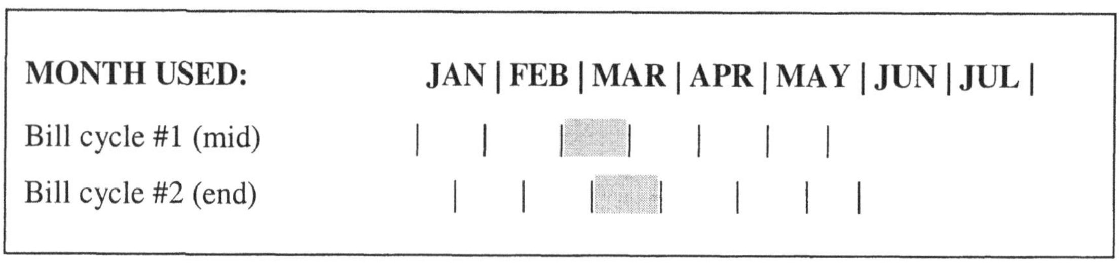

Figure 3.1 Monthly Billing Cycle

This lag effect is more pronounced in bi-monthly billing systems, where the water use reported in a billing period may include water use from up to two months prior. This effect is illustrated in Figure 3.2, showing a hypothetical four cycle bi-monthly billing system:

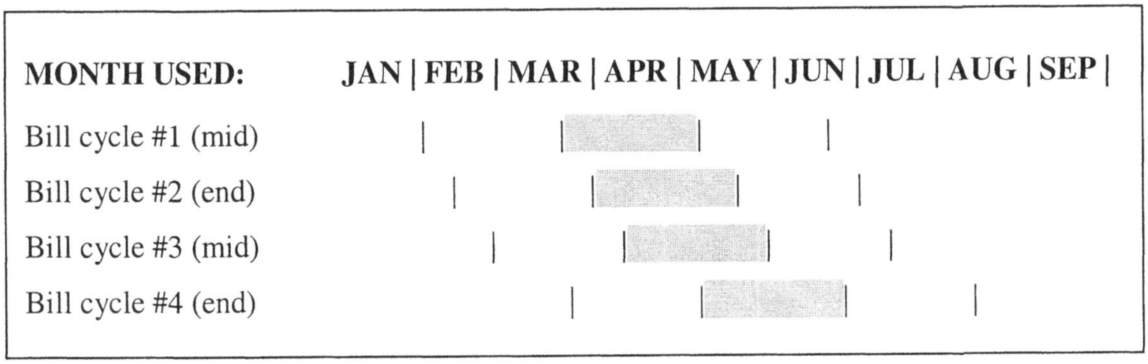

Figure 3.2 Bi-Monthly Billing Cycle

As shown, water used or delivered in May, is actually split among the four billing cycles, and is mixed with water used March, April and June. Chapter 5 provides a specification of a seasonal functional form that is continuous in calendar time (a Fourier series of sines and cosines). This form, coupled with the rolling nature of meter-read data, allows for a consistent estimate of the seasonal shape of mean water demand. Determining peak factors from this kind of data does require calibration to production data or system metering data.

CHAPTER 4
DATA AND SAMPLING

INTRODUCTION

In many uses of consumption data in distribution system modeling efforts, too little attention is paid to what can and cannot be done with available data. This research pays special attention to the strengths and limits of data typically available to water utilities. A sample design will be developed for historical water consumption data, data for the sample will be acquired from billing systems, and these data will be validated and cleaned. Spatial categories of interest will be identified and used to develop what scientific sampling terms "strata." In the following discussion the term "strata" and "category" will be used interchangeably. A stratified sample will then be designed where the sampling strata come from the desired spatial aggregation. Appropriate stratification weights will be developed to permit scientific inference from the sampled connections to the entire population of service connections. In this way, a formal basis is preserved for inference about the means and distributions surrounding spatial parameters.

The next task involves validation of the consumption data. "Validity" in scientific inference refers to the question of how a measure (consumption) relates to the object of interest (demand load). Due to line loss and measurement error, consumption data will imperfectly reflect flow requirements. To quantify this difference, consumption data for subareas need to be compared with corresponding demand (flow meter) data to quantify differences in sampled consumption and flow data. Completion of this task results in the creation of a consistent data source for the study (consumption time series containing data for all compared subareas). This chapter seeks to document the data strengths, limits, and workarounds.

Why Worry about Sampling?

Why should a distribution system modeler worry about developing a scientific sample? As one modeler in this project asked directly, "I have the 'actual' consumption data on each connection, why should I examine a subset of data?" Restated, if consumption data are available on the entire population of connections, why should they not be used? This entirely reasonable question deserves consideration by any modeler wondering why she should consider tackling the intimidating complexities of scientific sampling.

This report proffers three responses to the question of "Why sample?" The basic premise of scientific sampling is that inference on the population of connections can be made from a properly chosen sample. This is not to say that one need never collect data from the population of all possible demand nodes. To the contrary, data from the population are required to select a representative sample. Rather, the methods of scientific sampling impose method and rigor upon the practice of collecting data.

The second reason is probably the most important. Choosing a sample can be more cost-effective. Though manipulation of very large data sets has become increasing more cost-feasible, data validation remains cost intensive. If the modeler chooses to manipulate all potential service connections, then the time and effort allotted to data checking and validation will be quite limited on a per connection basis. Collecting consumption data across a sufficient number of years to embed weather fluctuations would likely be cost prohibitive. In short, a smaller dataset allows limited analytic resources to be spent more effectively.

The third reason relates to the depressing fact that any collection of consumption data will always be – due to corrupted data, other data attrition, and normal changes in the growth of customer connections—a sample of the future population of customers for which the modeler designs. Principles of scientific sampling allow the modeler to correct for errors in the data collection process so that inference to the future and possibly not yet observed population can be correctly made.

Though we run the risk of belaboring the point, readers are enjoined to consider the notion that the principles of scientific sampling can shed light on inferential risks, issues of data validity, and possible corrective measures that can be of great practical merit in applied distribution system modeling.

SAMPLE DESIGN, DATA COLLECTION, AND DATA VALIDATION

This research identified the following tasks toward developing a sample of water consumption that permits valid inference through time and space. Many of these steps may be followed currently in an implicit or ad hoc manner; stating these tasks formally can allow more rigorous review applied to each task.

a) **Design Consumption Data Sample-**in this task the analyst determines the sample size, defines the spatial data categories (strata), and, in the process, specifies the method for reweighing the sample to more closely approximate the population from which it is drawn (stratification weights). Each subtask will be discussed in turn.
   1. Select Spatial Data Categories (Strata) – The choice between strata usually depend upon the purpose for which the modeling is to be done. Near-term planning forecasts require land use strata; Calibration of consumption data to production subareas require production areas as strata. Selection of sample strata based on as fine a resolution as practical allows use of other broader strata at a latter date.
   2. Determine Sample Size and Sample Time Period. Selection of sample size is usually done for reasons of convenience. If one selects all current accounts with valid data, for example, then many sampling questions appear to be obviated. Similarly, the selection of the time period is often determined by the realistically important question of what data is easily available. We conclude noting that individual years of consumption data are very uninformative about fluctuations in demand load—wet years differ greatly from dry years. Without a reasonable sample of different realizations of weather, one is left with little inferential basis for how weather conditions affect demand load.
   3. Develop Stratification Weights-Samples typically differ from the target population, intentionally or otherwise. For example, only half of one strata are sampled, one will need to upweight the sample by two to reach conclusions about the population. In the example just cited, the stratification weight is two and is equal to the inverse of the probability of selection.

b) **Acquire Spatial Category Data (Strata)**-This task requires identifying the individual parcel, census tracts, or land use data that correspond to the data in the sample. This task often requires use of GIS or database analyses to prepare the spatial data in ways that can be matched to the consumption data sample.

c) **Acquire Raw Consumption Data** – This task requires acquiring the historical consumption data to match the selected sample of service connections. Historical consumption records are retained on different media at different water utilities (magnetic mainframe tapes, relational data bases, or other information system). Assessing these historical records can require use of either possibly arcane or certainly obtuse programming languages. The next task will struggle to make sense of the raw data collected in this task.

d) **Create Consistent Consumption Histories/Time Series (Data cleaning)**- This task involves combining connection consumption data to create a consistent time-series measure of consumption by connection. Some of these issues are discussed in the previous chapter. These issues include retaining records of residence change, integrating data on meter replacement, proper handling of estimated or missing meter reads, and methods to identify wrongly entered or mismeasured consumption records.

e) **Populate Consumption Histories Data Sample with Spatial Category (Strata) Data (Historic and Existing)**-This task involves connecting spatial indicators (land-use categories, production sub-areas, customer class. The progress made by geographical information systems in many water utilities affords new opportunities for integrating existing spatial data with existing consumption records.

f) **Design Sub-Area Water Delivery Data Sample**-in this task, the sub-area consumption measures are compared to sub-area flow meter measurement. Production flow measurement in sub-areas of the service area often occurs as part of distribution design studies; By comparing these existing measures against comparable sub-area consumption data, a means for estimating differences between the two exists.

g) **Develop Data Documentation**-Good documentation of the steps taken in the sampling design, data collection, and data validation process allow for easier reenactment when errors of implementation occur or new data become available. Replicablity is also a cornerstone of the scientific method.

Sample Design

One of the first choices in the sample design for this research was the selection of sampling strata. Three possible types of sample strata—representing three types of spatial categories—were considered by this project:

- Sample Strata 1 – Production Groupings (13 subareas)
- Sample Strata 2 – Land Use Classifications
- Sample Strata 3 – Customer Class Groupings

One would want to use different strata for different purposes[1]. The production groupings would be used for validation of consumption data against production measures. The land use classifications would be used for predictive purposes—known land use changes may be some of the most certain predictors of tomorrow's spatial demand. The customer class groupings would be used to estimate the highest resolution consumption models. From these models, one may be able to derive predictable trends in consumptive end uses. The approach taken was to construct one consumption sample back through time that can be viewed through three different spatial lens.

Customer class was selected as the first level of sampling strata for several reasons:

- Customers within a customer class tend to be similar
    - Similar customers will require similar models
    - Dissimilar customers require different models
- Water rates and charges differ by customer class
    - Rates and changes can affect consumptive patterns

A primary stratification by customer class also permits consideration of different additional sampling schemes for each class. The use of different sampling strategies for different customer classes allows this research to illustrate different approaches. The "best" sampling strategy always depends on the particulars—the quality of alternative measures of spatial indicators, the intended application, and the physical model of what causes demand to vary.

Single-family customers, for example, have consumptive patterns that vary strongly with climate zone and lot size. The Department of Water and Power customer data classifies all connections into one of three temperature zones and one of five categories of lot size for purposes of their block rate structure. The various combinations of temperature zones and lot sizes (3 zones X 5 lot sizes = 15 strata) represent another layer of stratification. This very fine level of stratification allows recombination into broader categories—production subareas, for example—as the need requires. Since we would like resolution within each stratum, the sample size requirement (approximately 300) applies to each stratum. Though this makes for a large sample of connections (approximately 15 strata X 300 sampled = 4,500), this sample can provide good geographical inference to other potential strata.

Multiple family customers tend to use a larger proportion of their water use for indoor end uses. A key determinate of the water use of a multiple family complex is the number of dwelling units. The number of dwelling units would be very suitable for a secondary level of stratification. (This research used another more commonly available stratification measure—meter size.)

---

[1] One stratification variable ruled out for use was consumption itself. Use of the dependent variable for stratification vastly complicates the inferential task. Selecting low, medium, and high water users as strata, for example, results in a "censored" dependent variable that requires "censored data models". The dependent variable for low water users would likely not have any of the highest consumption values; These have been censorerd by the self-stratification. See Wooldridge (2002) for a discussion of these models, especially section 17.8.2 on the use of weighted estimators to account for stratification.

Industrial, institutional, and commercial customers vary greatly and exhibit little coherence. Even more complicated sampling schemes based on North American Industry Classification System (NAICS superceded the SIC system in 1997). Given the greater degree of heterogeneity a greater proportion of these accounts will need to be sampled. Indeed, it is exactly in the case of high heterogeneity where the argument for developing as large a sample as practical is strongest. Contrariwise, if all customers within a class were exactly identical to one another—that is, homogenous—then a sample size of one would be perfectly sufficient to allow perfect inference.

Enactment of the Stratified Sample

To divide the entire sample of customers into customer classes, a field from the billing system (rate code) was used. Fields used to generate customer bills tend to have higher levels of reliability, given the opportunity for customer review and validation.

Single Family Sample

The single-family sample was created by selecting those connections with this classification (the billing system contain a rate code for single family connections, RC=30), approximately 396,120 connections. The "water code" field was used as the basis for stratification. Water code is used in the billing system to define the block in a single family customer's block rate—customers in the high temperature zones or having large lot sizes are allocated a slightly larger first block. Water code defines five lot size categories and three temperature zones, resulting in a total of 15 (5X3) possible strata. Approximately 1.25 percent of the connections within each stratum were randomly drawn, resulting in a sample size of approximately 5000 service connections.

Multiple Family Sample

The multiple family sample was created by selecting based upon rate code (RC=33). Since multiple family accounts are not billed similarly to single-family customer—i.e., using lot size/temperature zone block rates—a different approach was taken. Meter size was used as the primary basis for stratification. In addition to this, the sample used a geographical indicator (service district) to control for geographic area.

Since each multiple family meter can represent many dwelling units, it is important to control for size. The size of a water meter is correlated with the volumetric consumption of a multiple family complex. Using meter size as a stratification variable is one step in this direction. The proportion randomly selected from each meter size was also designed to vary roughly according to the volume of water accounted for by each meter size. Thus, one percent of the smallest meter size (one inch) was randomly selected while all of the largest meter size were selected. In the multiple family sample, we wanted to oversample larger complexes to increase the resolution of the sample. Within each of DWP districts, the probability of (random) selection varies proportionally to the volume of water accounted by each meter size, as depicted in Table 4.1

Table 4.1
Probability of Selection based on Meter Size Strata for Multiple Family Sample Design

Meter Size	Probability of Selection (p)	Post-Stratification Weight (p^{-1})
.75-1 inch	1%	100
1.25 inches	3%	33.33
2 inch	8%	12.5
3 inch	25%	4
4 inch	50%	2
≥6 inch	100%	1

The resulting random sample contained approximately 3,000 connections, representing water consumption for about 4,700 different customer accounts.

The commercial industrial and institutional customer class was sampled using district as the only stratum and contained approximately 5000 connections, about half of the total population of connections.

The enactment of the randomization can be understood through the following logic. A random number is generated for each stratum from a uniform distribution between zero and one. Thus, each connection is associated with a fraction between zero and one. The connections within each stratum are sorted by the random number. A 50% probability of selection would then involve taking all connections whose random number is less than 0.5. Similarly, a quote sample of 5000 would involve taking the first 5,000 connections after the sort by random number.

CHAPTER 5
SPECIFICATION OF THE STATISTICAL MODEL

INTRODUCTION

There are three issues that need to be addressed in specifying a statistical model of water consumption.

- Issue 1: Why model?
- Issue 2: Specify the dependent variable
- Issue 3: Specify the model form

Each issue will be addressed in turn.

Issue 1: Why model?

It is important to understand the reasons for developing a formal consumption model in order to justify the effort required. Another way to think of this is that if one cannot find an answer to the question "Why model?" then the task of model specification need not be undertaken. Three reasons will be provided here.

The very messy nature of consumption data provides the first reason: A model can provide a means of extracting the signal from random noise in the consumption data. The regression models can apportion the variation in the consumption data into two parts: the portion that can be explained by the included variables and the part that cannot. Furthermore, estimated models can also be extended to identify problematic data. In this research, statistical models were used to attach a measure of "suspiciousness" to each observation. (If an observation differs greatly from what the model expects to see, it is deemed "suspicious.")

The second reason for developing a formal model of consumption lies in the scientific inference that models permit. The statistical models formulated in this chapter can be used to develop formal answers to the follow types of questions:

- What is mean demand for a customer class?
- How can demand vary?
  - across customers
  - through the year, or
  - as a function of weather variation
- What is a 99.9% level of demand load?

The third reason for developing a formal model of consumption is that models can provide useful answers.

Issue 2: Specify the dependent variable

Consider a general form of a consumption model expressed in functional notation:

$$\text{Consumption} = f(\text{Time and Space})$$

The first step in specifying a model of consumption is to settle on a specification for the consumption measure, termed the dependent variable or the Left-Hand Side (LHS) of the equation. Alternatives include:

- Quantity (gallons per day) = **f**(")
- Quantity/Area (Acre-feet/Acre)=**g**(")

Note that the specification of the LHS can imply the need of a different functional explanation (**f** versus **g**, above.) The models used in this report convert the original measure of consumption (hundred cubic feet per billing period) to gallons-per-day (gpd). This is done for two reasons. First, it ensures time-consistency. Billing periods tend to vary in length as a natural part of the process of meter reading. Additionally this conversion allows for comparisons across billing periods of different lengths—monthly, bimonthly, or quarterly. Secondly, this specification still allows inference on a per acre basis; the consumption is forecasted and then combined the area measure.

Issue 3: Specify the model form

The right hand side (RHS) of the equation contains the variables used to explain how water consumption varies through time and space. The model form can and should depend on the application. More than one model form will be considered here. In fact, a range of model forms will be examined beginning with the simplest and working toward the more complicated. In this way, simple concepts such as the distinction between model variables and model parameters can be made while also explaining their relationship to reality.

Simple Model 1 - Consumption as a function of time

Consider a model that specifies the consumption quantity as a linear function of time:

$$\text{Quantity} = f(\text{Time})$$
$$Q_t = \beta_0 + \beta_1 \cdot \text{Trend}_t + \varepsilon_t$$

As illustrated in the Figure 5.1, the function **f** is a line with intercept β_0 and slope β_1. The variable Trend_t is an index of time: an annual trend would increase in value by one each year. Note that the parameter ε_t represents the *"error term"*: the vertical distance from any observation to the line defined by **f**.

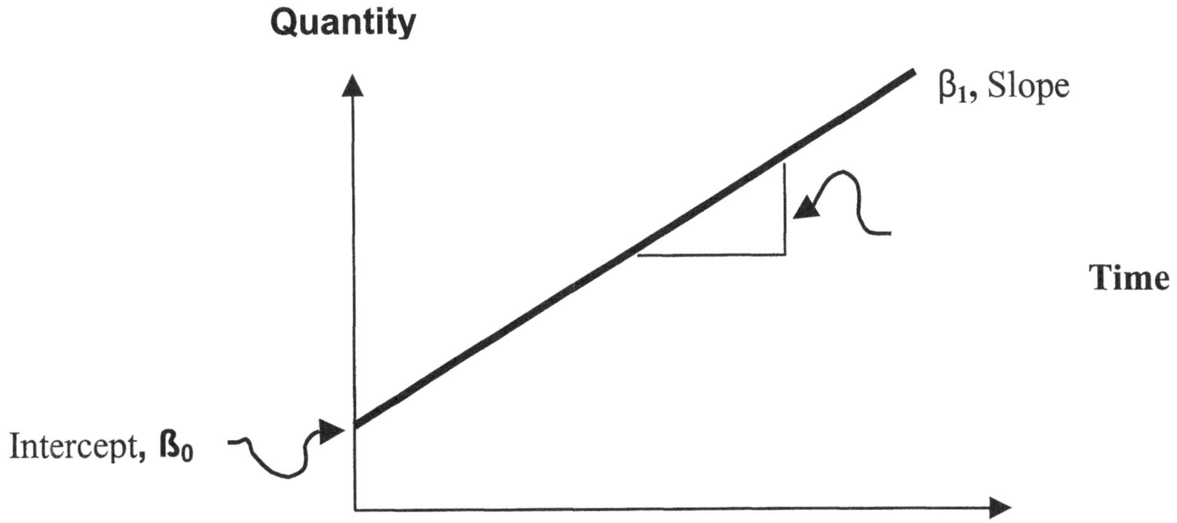

Figure 5.1 Simple Model 1 – Consumption as a function of time

In a similar fashion, one can specify a model that is only a function of spatial parameters.

Simple Model 2 - Consumption as a function of space

Consider next a model that specifies the consumption quantity solely as a function of spatial characteristics. A model of average consumption in two different service areas might look like this:

$$\text{Quantity} = \text{f(Space)}$$

$$Q_i = \mu_1 \cdot \text{Area}_1 + \mu_2 \cdot \text{Area}_2 + \varepsilon_i$$

As illustrated in Figure 5.2 below, the function **f** is represented by two levels of average consumption, μ_1 and μ_2. The variable Area_1 takes on the value 1 if the consumption occurs in Area 1 and is zero otherwise. The variable Area_2 is a similar indicator variable constructed in a similar fashion. A regression of consumption on spatial indicator variables will produce estimates of the average of consumption in the indicated area.

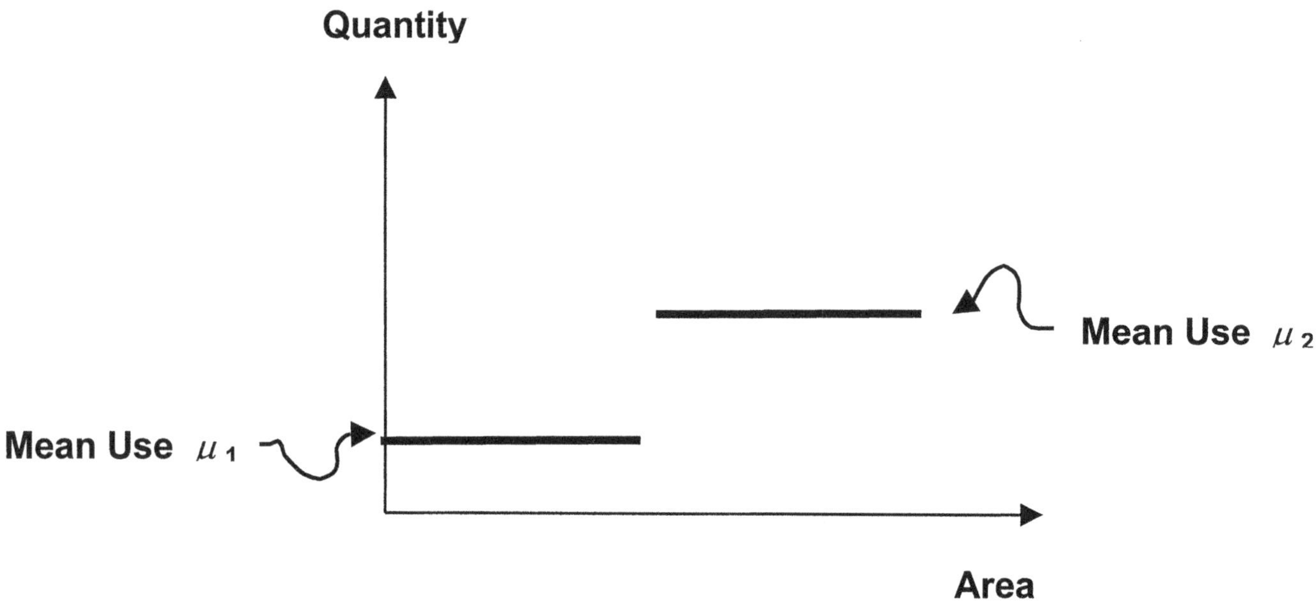

Figure 5.2 Simple Model 2 - Consumption as a function of space

Note that this second model requires an indicator variable for each spatial area described. Additional variables can be added to include additional effects. Though we next move to more general model forms, it is important to understand the distinction between spatially-invariant time variables (as illustrated by the Trend variable of the first model) and the time-invariant spatial variables (illustrated by the Area indicator variables of the second model.

Simple Model 3 - Consumption as a function of things that vary in time and space

A model that specifies consumption as a function of things that vary either in time, or in space, or in both might include a list of variables thought to influence consumption: measures of the time of year (season), weather (temperature and rain), or other economic forces (the price of water.) A model of household consumption might also allow for household specific differences and could look like this:

$$\text{Quantity} = f(\text{Season}, \text{Temp}, \text{Rain}, \text{Price})$$
$$Q_t = \mu_i + \beta_1 \cdot Season_t + \beta_2 \text{AirTemp}_t + \beta_3 \cdot \text{Rain}_t + \beta_4 \cdot \text{Price}_t + \varepsilon_t$$

Here, each slope coefficient β_i represents a measure of the empirical relationship between the driver variable and water consumption. The graph below illustrates such a linear relationship. If rainfall were the variable in question, it's slope coefficient β_3 could be interpreted as the amount of change produced in consumption by a one unit change in rainfall.

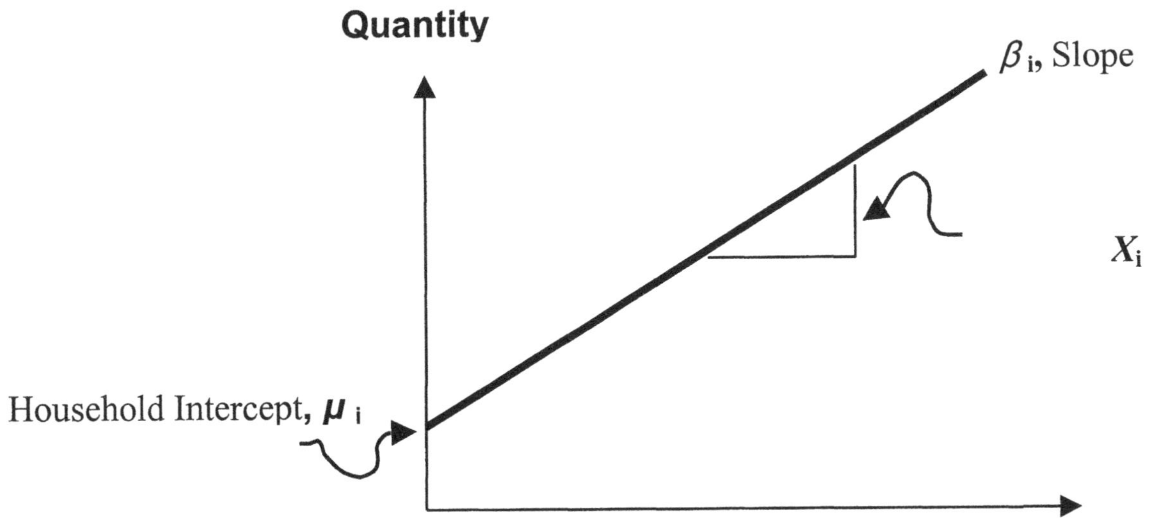

Figure 5.3 Simple Model 3 - Consumption as a function of space and time

MODEL SPECIFICATION

This section specifies a function that has several unique features. First, it models seasonal and climatic effects as continuous (as opposed to discrete monthly, semi-annual, or annual) function of time. Thus, the seasonal component in the water consumption model can be specified on a continuous basis, then aggregated to a level comparable to measured water use (e.g. bimonthly). Second, the climatic component is specified in difference form as a similar continuous function of time. The climate measures are thereby made independent of the seasonal component. Third, the model permits interactions of the seasonal component and the climatic component. Thus, the season-specific response of water demand can be specific to the season of the year.

The general form of the model is:

$$Q_t = \mu_i + S_t + C_t + T_t \qquad \textbf{(5.1)}$$

where $\boldsymbol{Q_t}$ is the quantity of water consumption within time t, the parameter $\boldsymbol{\mu_i}$ represents mean water consumption per customer i, $\boldsymbol{S_t}$ is a seasonal component, $\boldsymbol{C_t}$ is a climatic component, $\boldsymbol{T_{,t}}$ represents a component of any other time-varying variables. Each of these components is described below.

Seasonal Component

A monthly seasonal component can_be formed using monthly dummy variables to represent a seasonal step function. Equivalently, one may form a combination of sine and cosine terms in a Fourier series to define the seasonal component as a continuous function of time.[1] The following harmonics are defined for a given day T, ignoring the slight complication of leap years:

$$S_t \equiv \sum_{1}^{6}\left[\beta_{i,j} \cdot \sin\left(\frac{2\pi \cdot jT}{365}\right) + \beta_{i,j} \cdot \cos\left(\frac{2\pi \cdot jT}{365}\right)\right] = Z \cdot \beta_S \qquad \textbf{(5.2)}$$

where T = (1,...365) and j represents the frequency of each harmonic.[2] Because the lower frequencies tend to explain most of the seasonal fluctuation, the higher frequencies can often be omitted with little predictive loss.

To compute the seasonal component one simply sums the multiplication of the seasonal coefficient with its respective value. This number will explain how much demand has changed due to the total seasonal fluctuation.

Climatic Component

The model incorporates two types of climate measures into the climatic component– maximum daily air temperature and rainfall.[3] The measures of temperature and rainfall are then logarithmically transformed to yield:

$$R_t \equiv \ln\left[1 + \sum_{t=T}^{T_d} Rain_t\right], A_t \equiv \ln\left[\sum_{t=T}^{T_d} \frac{AirTemp_t}{d}\right] \qquad \textbf{(5.3)}$$

where d is the number of days in the meter read. (For daily models, d would take on the value of one.) Because weather exhibits strong seasonal patterns, climatic measures are strongly correlated with the seasonal measures. In addition, the occurrence of rainfall can reduce expected air temperatures. To obtain valid estimates of a constant seasonal effect, the seasonal component is removed from the climatic measures by construction.

---

[1] The use of a harmonic representation for a seasonal component in a regression context dates back to Hannan (1960). Jorgenson (1964) extended these results to include least squares estimation of both trend and seasonal components.

[2] If measures of water consumption are available on a daily basis, the harmonics defined by Equation 2 can be directly applied. When measures of water demand are only observed on a monthly or bimonthly basis, two steps must be taken to ensure comparability. First, water consumption should be divided by the number of days in the reading period to give a measure of average daily use. Otherwise, the estimated seasonal component will be distorted by the differing number of days in a reading period.

[3] Specifically it uses the maximum daily air temperature and the total daily precipitation at four geographically diverse weather stations within each of the four service districts: the Los Angeles Civic Center (District 1-City Center), Santa Monica Pier (District 2-West Los Angeles), Canoga Park (District 3-San Fernando Valley), and Long Beach (District 4-Harbor) stations.

Specifically, climatic measures are constructed as a departure from their "normal" or expected value at a given time of the year. The expected value for rainfall during the year, for example, is derived from regression against the seasonal harmonics. The expected value of the climatic measures ($\hat{\boldsymbol{A}}=\boldsymbol{Z}\hat{\boldsymbol{\alpha}}$) is subtracted from the original climatic measures:

$$C_t \equiv (R_t - \hat{R}_t)\cdot\beta_R + (A_t - \hat{A}_t)\cdot\beta_A \qquad \textbf{(5.4)}$$

The climatic measures in this deviation-from-mean form are thereby separated from the constant seasonal effect.[4] Thus, the seasonal component of the model captures all constant seasonal effects, as it should, even if these constant effects are due to normal climatic conditions. The remaining climate measures capture the effect of climate departing from its normal pattern.

The model can also specify a richer texture in the temporal effect of climate than the usual fixed contemporaneous effect. Seasonally-varying climatic effects can be created by interacting the climatic measures with the harmonic terms. In addition, the measures can be constructed to detect lagged effects of climate, such as the effect of rainfall one month ago on this month's water demand.

Stochastic Effects

To complete the model, we must account for the fact that not every data point will lie on the plane defined by **Equation (5.1)**. This fundamental characteristic of all systematic models can impose large inferential costs if ignored. Misspecification of this "error component" can lead to inefficient estimation of the coefficients defining the systematic forces, incorrect estimates of coefficient standard errors, and an invalid basis for inference about forecast uncertainty.

The specification of the error component involves defining what departures from pure randomness are allowed. What is the functional form of model error? Just as the model of systematic forces can be thought of as an estimate of a function for the "mean" or expected value, so too can a model be developed to explain departures from the mean—i.e., a "variance function" If the vertical distance from any observation to the plane defined by **Equation (5.1)** is the quantity ε, then the error component is added to **Equation (5.1)**:

$$\ln Use = \mathbf{f}(\mu_i, \mathbf{S}_t, \mathbf{C}_t, \mathbf{T}_t) + \varepsilon \qquad \textbf{(5.5)}$$

[4] The logarithmic transformation of the original climate variable implies that the seasonal mean climate effect is a geometric mean. Because the model is estimated on the logarithmic scale the departure-from-mean climatic effects would be more accurately termed departure-from-median. See Goldberger, A.S. (1968) "The Interpretation and Estimation of Cobb-Douglas Functions," *Econometrica*, vol. 35, p. 464-472.

The error structure is assumed to be of the form:

$$\varepsilon_{it} \sim N(0,\sigma_{\varepsilon}^{2}) \tag{5.6}$$

This is certainly not the only possible specification of the stochastic effects in the model. The model form specified by the above two equations is termed a "fixed-effects" model due the existence of customer-specific intercepts μ_i that, by construct, are fixed. The individual component μ represents the effects of unmeasured household characteristics on household water use. An example of such an unmeasured characteristic might be the water use behavior of household members. This effect is constant over the estimation period.

An alternative "random-effects" specification would assert that the customer-specific effects are random, instead of fixed, coming from a probability distribution. Such a specification would place the μ_i in the error term:

$$\begin{aligned} &\varepsilon_{it} = \mu_i + \xi_{it} \\ &\textit{where} \\ &\mu_i \sim N(0,\sigma_{\mu}^{2}) \\ &\xi_{it} \sim N(0,\sigma_{\xi}^{2}) \end{aligned} \tag{5.7}$$

The second component ξ represents random error. If the μ and ξ are independent, the error variance can be decomposed into two components:

$$\sigma_{\varepsilon}^{2} = T \cdot \sigma_{\mu}^{2} + \sigma_{\xi}^{2} \tag{5.8}$$

This "random-effects" model specification is accordingly called an error components or variance components model.

The choice between fixed or random-effects specification can be guided by empirical tests. Hausman's classic paper on specification tests (1976) used exactly this example as one illustration of specification tests. This report recommends that fixed- effects models are a smart place to begin specification for several reasons[5]:

(1) Fixed-effects models are relatively easy to understand and explain.
(2) Fixed-effects specification is relatively simple to enact (they can be enacted by either including a customer-specific dummy variable in standard regression packages, or implicitly by subtracting customer-specific means from all variables used in the regression.)
(3) The fixed-effects specification is robust across a wider range of distributional assumptions (the assumption of independence between the error term and the systematic portion of the model is often questionable. This violation of assumptions

---

[5] Standard references for time series cross sectional models include Hsiao (1986) *Analysis of Panel Data* and Wooldridge (2002) *Econometric Analysis of Cross Section and Panel Data.*

produces biased coefficient estimates in standard random-effects model; fixed-effects estimator remain consistent, even if inefficient[6].)

Though many distribution system modelers may have little experience with time-series cross-section modeling, many have had some exposure to regression analysis. Fixed-effects models can be enacted with any regression analysis software and can be interpreted in a standard analysis of covariance framework. The next chapter illustrates the estimates of these models applied to the customer-class samples of Chapter 4.

---

[6] Mundlak (1978) demonstrated that the estimator that allows for correlation between the error term and model variables (a random coefficients estimator) was asymptotically equivalent to the fixed-effects estimator. Hsiao (1986) notes conditions where this does not occur.

CHAPTER 6
ESTIMATED CONSUMPTION MODELS

ESTIMATION RESULTS

This section presents the statistical estimates of water consumption models for the different customer classes. These models are described as "water consumption" models rather than "water demand" models because they do not model the effect of price on customer water consumption. This requires a more involved structural model, especially when handling the effect of increasing block water rate structure.[1]

Estimated Single Family Residential Water Consumption Model

Table 6.1 presents the estimation results for a model of single family water consumption in the City of Los Angeles. The sample includes 7,944 unique customers and 214,723 unique meter read consumption observations over the period January 1992-August 2000. The independent variables 1 to 10—made up of the sines and cosines of the Fourier series described in Equation 2—are used to depict the seasonal shape of water demand. The predicted seasonal effect (that is, $\hat{S}_t = Z \cdot \hat{\beta}_S$) is the shape of demand in a normal weather year. This seasonal shape is important in that it represents the point of departure for the estimated climate effects (expressed as departure from normal).

The estimated weather effect is specified in "departure-from-normal" form. Variable 13 is the departure of monthly temperature from the average temperature for that month in the season. (Average seasonal temperature is derived from a regression of daily temperature on the seasonal harmonics.) Rainfall is treated in an analogous fashion (Variable 16). One period lagged rainfall deviation is also included in the model (Variable 17). The reader should also note that the contemporaneous weather effect is interacted with the harmonics to capture any seasonal shape to both the rainfall (Variables 14 and 15) and the temperature (Variables 11 and 12) elasticities. Thus, departures of rainfall from normal produce the largest percentage effect in the spring growing season. Similarly, a departure of temperature from normal produces a larger effect upon demand in the summer than in the winter. The constant term (18) describes the estimated mean customer-specific intercept for this equation. The estimated standard error of the customer-specific intercepts ($\hat{\sigma}_{\mu_i}$) is 0.706.

---

[1] Technically, block rates introduce a causal complication because the price of water itself is a function of the amount of water consumption. Economists simultaneously estimate two equations—one for price and one for quantity demanded—to attempt to untangle the causality between price and quantity. See Billings (1982), Jones and Morris (1984), and Hanemann (1984) for an explication of these issues. This is known as the "identification" problem in econometrics. This research does not specify a causal/structural model for how price, income and conservation programs affect water demand.

Table 6.1
Single Family Residential Water Consumption Model
Dependent Variable: Log Average Daily Metered Water Consumption
(in gallons per day)

Independent Variable	**Coefficient**	**Std. Error**
1. First Sine harmonic, 12 month (annual) frequency	-0.1654	0.00117
2. First Cosine harmonic, 12 month (annual) frequency	-0.2898	0.00115
3. Second Sine harmonic, 6 month frequency	-0.0118	0.00131
4. Second Cosine harmonic, 6 month frequency	0.00157	0.00125
5. Third Sine harmonic, 4 month frequency	-0.0093	0.00165
6. Third Cosine 4month frequency	0.01299	0.00156
7. Fourth Sine harmonic, 3 month frequency	0.01311	0.00246
8. Fourth Cosine, 3 month frequency	0.00365	0.00247
9. Fifth Sine harmonic	-0.0071	0.00522
10. Fifth Cosine harmonic	-0.0376	0.00527
11. Interaction of contemporaneous temperature with annual sine harmonic	0.39253	0.03207
12. Interaction of contemporaneous temperature with annual cosine harmonic	-0.0446	0.03921
13. Deviation from ln(31 or 61 day moving average of maximum daily air temperature)	0.81103	0.02531
14. Interaction of contemporaneous rain with annual sine harmonic	-0.0489	0.00268
15. Interaction of contemporaneous rain with annual cosine harmonic	-0.0235	0.00324
16. Deviation from ln(31 or 61 day moving sum of rainfall+1)	-0.0672	0.00234
17. Bi-monthly lag from rain deviation	-0.0301	0.00137
18. Constant (Mean intercept)	5.78905	0.00077
Number of observations	214,723	
Number of customer accounts	7,944	
Standard Error of Individual Constant Terms	0.70564	
Standard Error of White Noise Error	0.32573	
Time period	January 1992- August 2000	

Estimated Multiple Family Residential Water Consumption Model

Table 6.2 presents the estimated model for multiple family customers. Similar in form to the model for single family customers, fewer terms are needed to capture both the seasonal effect and the effect of weather on water consumption.

Table 6.2
Multiple Family Residential Water Consumption Model
Dependent Variable: Log Average Daily Metered Water Consumption
(in gallons per day)

Independent Variable	Coefficient	Std. Error
1. First Sine harmonic, 12 month (annual) frequency	-0.0695	0.00107
2. First Cosine harmonic, 12 month (annual) frequency	-0.0726	0.00099
3. Second Sine harmonic, 6 month frequency	0.00555	0.00108
4. Second Cosine harmonic, 6 month frequency	-0.0047	0.00111
5. Fourth Sine harmonic, 3 month frequency	0.0038	0.00165
6. Fourth Cosine, 3 month frequency	-0.0019	0.00167
7. Fifth Cosine harmonic	-0.0032	0.00216
8. Interaction of contemporaneous temperature with annual sine harmonic	-0.0581	0.02762
9. Interaction of contemporaneous temperature with annual cosine harmonic	-0.2723	0.02927
10. Deviation from ln(31 or 61 day moving average of maximum daily air temperature)	0.2577	0.02372
11. Interaction of contemporaneous rain with annual sine harmonic	-0.0083	0.00259
12. Deviation from ln(31 or 61 day moving sum of rainfall+1)	-0.0146	0.00183
13. one period lag from rain deviation	-0.0022	0.00142
14. Constant (Mean intercept)	7.90359	0.00295
Number of observations	149,557	
Number of customer accounts	4,709	
Standard Error of Individual Constant Terms	1.4551	
Standard Error of White Noise Error	0.24908	
Time period	January 1992- August 2000	

Estimated Commercial Industrial Institutional Water Consumption Model

Table 6.3 presents the estimated model for Commercial, Industrial, and Institutional customers. Similar in form to the previous models, this more closely resembles the multiple family model. The seasonal effect is relatively flat and the effect of weather on water consumption is smaller in magnitude than single family consumption.

Table 6.3
Commercial Industrial Institutional Water Consumption Model
Dependent Variable: Log Average Daily Metered Water Consumption
(in gallons per day)

Independent Variable	Coefficient	Std. Error
1. First Sine harmonic, 12 month (annual) frequency	-0.0705	0.00183
2. First Cosine harmonic, 12 month (annual) frequency	-0.089	0.00174
3. Second Sine harmonic, 6 month frequency	0.00016	0.00178
4. Second Cosine harmonic, 6 month frequency	-0.0219	0.00176
5. Third Sine harmonic, 4 month frequency	0.00065	0.00183
6. Third Cosine 4month frequency	-0.0105	0.0019
7. Fourth Sine harmonic, 3 month frequency	0.01176	0.00207
8. Fourth Cosine, 3 month frequency	-0.0056	0.00213
9. Fifth Sine harmonic	0.01007	0.00239
10. Fifth Cosine harmonic	-0.0032	0.00239
11. Deviation from ln(31 or 61 day moving average of maximum daily air temperature)	0.35581	0.0357
12. Interaction of contemporaneous rain with annual sine harmonic	-0.0144	0.00509
13. Deviation from *ln* (31 or 61 day moving sum of rainfall+1)	-0.0188	0.0045
14. one period lag from rain deviation	-0.0119	0.00255
15. Constant (Mean intercept)	6.01554	0.00808
Number of observations	318,600	
Number of customer accounts	9,740	
Standard Error of Individual Constant Terms	1.5119	
Standard Error of White Noise Error	0.61982	
Time period	January 1992- August 2000	

Comparison of the Models

To illustrate these models, several graphs will be compared the estimated effects of the two residential consumption models. Figure 6.1 plots the seasonal pattern of single- family water consumption ($Z \cdot \widehat{\beta}_{S,SF}$) against the average daily air temperature (for the LA Civic Center Station over the period 1948-2000) through the days of the year. The seasonal pattern of multiple family water consumption ($Z \cdot \widehat{\beta}_{S,MF}$) is plotted in the next Figure 6.2. The multiple family seasonal pattern is smaller in magnitude than the single family seasonal pattern. This is expected since single family customers tend to have a larger proportion of their water consumption taken up by outdoor water end uses.

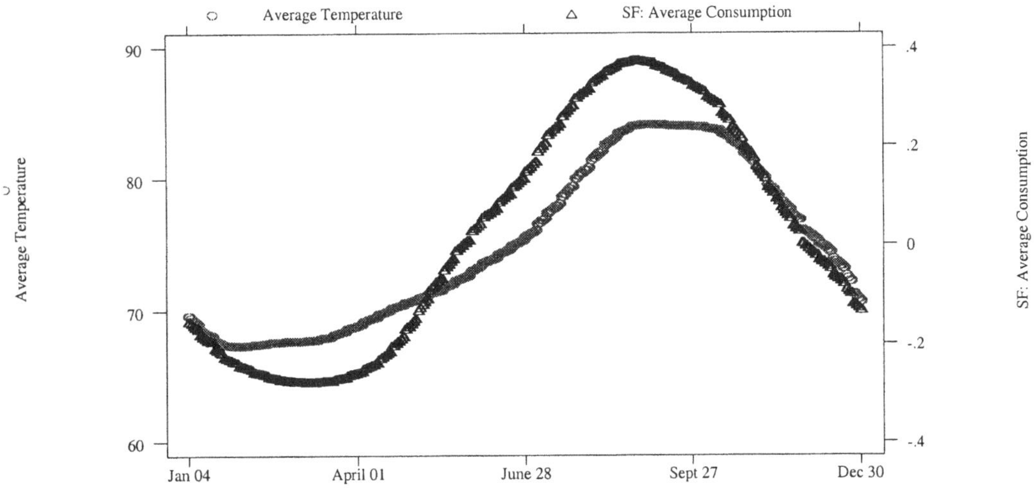

Figure 6.1 Single Family Seasonal Temperature and Seasonal Effect, Jan1 – Dec31

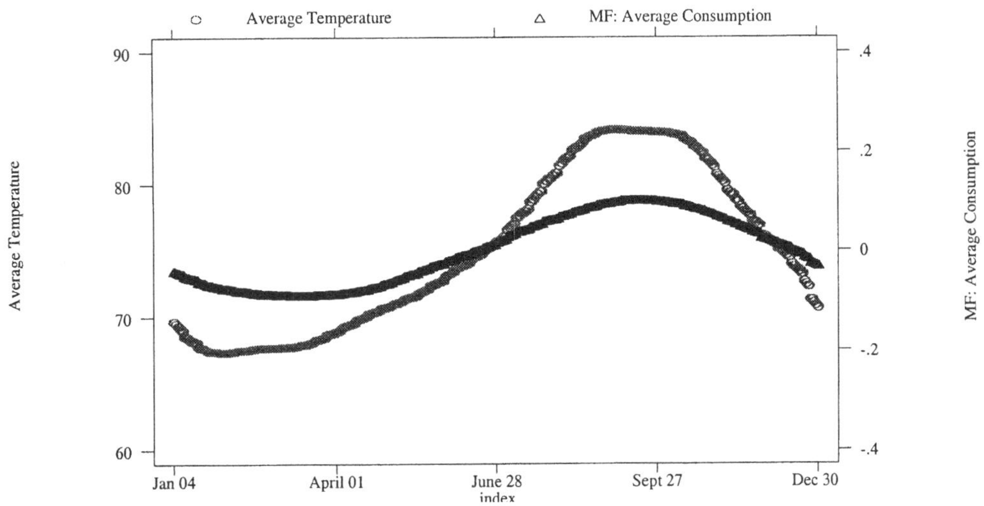

Figure 6.2 Multiple Family Seasonal Temperature and Seasonal Effect, Jan1 – Dec31

The response of water consumption to weather variations also differs by customer class. As expected, single family water consumption is more sensitive to rainfall and temperature (mean elasticities of $\widehat{\beta}_{R,SF} \approx -0.067$ and $\widehat{\beta}_{A,SF} \approx 0.811$) than the multiple family water consumption (mean elasticities of $\widehat{\beta}_{R,MF} \approx -0.015$ and $\widehat{\beta}_{A,MF} \approx 0.258$). These elasticities also vary through the year as plotted below in Figure 6.3. The rainfall elasticity at any point in the year can be derived from the partial derivative with respect to rainfall:

$$\beta_{R,t} = \widehat{\beta}_R + \widehat{\beta}_{R,\sin 1} \cdot Sin_1 + \widehat{\beta}_{R,\cos 1} \cdot Cos_1 \approx \text{-}0.067 \text{ - } 0.0489 \cdot Sin_1 - .0235 \cdot Cos_1$$

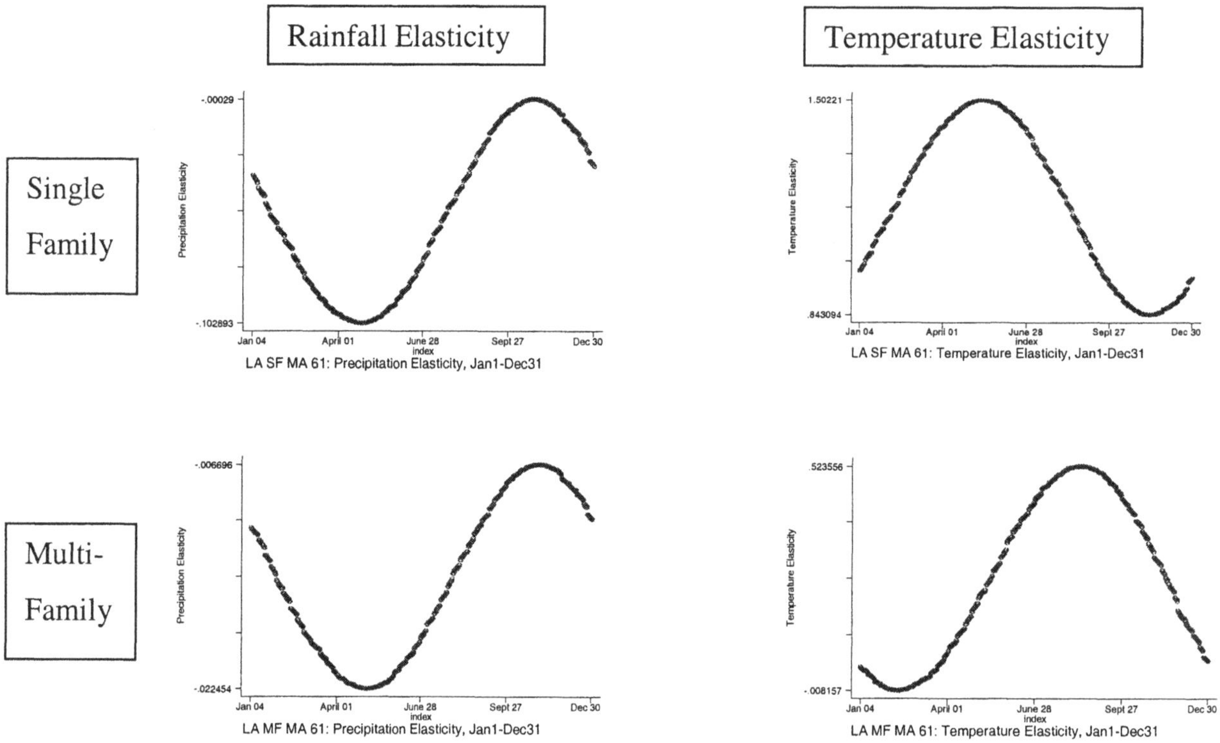

Figure 6.3 Comparison of Rainfall and Temperature Elasticities through the Year

Extensions of the Models

The basic models of water consumption by customer class can also be extended in useful ways.

Geographically Specific Models

The estimated customer-specific effects can graphed by smaller geographic areas and interpreted as a normal year baseline use. Figure 6.4 graphs a detailed look at the spatial demand distribution of the weather normalized average water consumption in gallons per day for single family customers broken out by production subarea. (The total of all single family customers in the DWP service area is graphed on the bottom right.) Different areas can certainly be characterized by different demand distributions. Breakdowns by subareas specific to measured production also allows for more detailed analyses of unaccounted for water that compare production measures to consumption measures. Needless to say, this application can have large revenue implications for most water utilities.

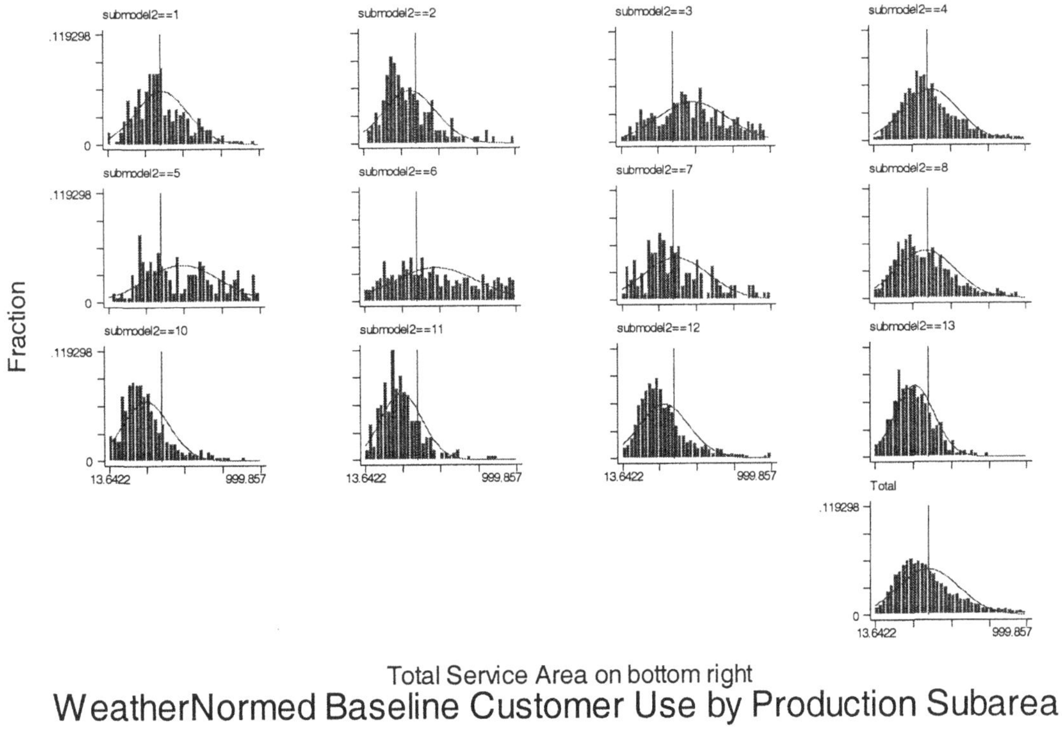

Figure 6.4 Single Family Use by Production Sub-area

Alternatively, the entire model can be re-estimated for smaller geographic areas to test for other important differences. Appendix B presents estimates of the single family model for each of 13 subarea production zones. The characteristic curves for the seasonal shape of average daily demand and weather elasticities differ substantially across the service area. Capturing these differences in water consumption can greatly improve the precision of model predictions.

Predicted Weather Effects

The models can also be used to generate the predicted effect of weather on water consumption. Substituting actual realizations for rainfall and air temperature into the estimated weather effect yields the models estimate of the effect of weather for the time period in question:

$$\hat{C}_t = (R_t - \widehat{R}_t) \cdot \hat{\beta}_R + (A_t - \widehat{A}_t) \cdot \hat{\beta}_A$$

This can also be expressed in percentage form:

$$C_t\% = \left[\frac{\exp(\hat{Q}_t) - \exp(\hat{Q}_t - \hat{C}_t)}{\exp(\hat{Q}_t)} \right]$$

Enacting this the predicted weather effect, for example, was done to identify the days that weather historically had the greatest effect on water consumption. (The maximal weather-induced effect in our sample occurred on March 31, 1997, raising single family demand by about 37 percent.) The estimated weather effect can also be extended further back in time than a consumption sample itself would allow. This "backcasting" exercise can be used to identify historical weather-induced peaking events and develop the corresponding simulated consumption.

Improvements to the Models

The single greatest improvement to the models presented in this chapter may be inclusion of a surrogate for the amount of outdoor water use. To the extent that customers are similar, then the average seasonal effect specified in the models above suffices. (Arguably, the approach works well enough for similar kinds of single family customers in similar temperature zones.) The seasonal shape of demand is strongly driven by outdoor end uses. Since several water utilities have begun programs to integrate aerial measures of landscape area into their geographical information systems, incorporating the effect of landscape end uses could go a long way toward improving the ability of consumption models to better predict customer peaking behavior.

CHAPTER 7
MODEL APPLICATION TO DISTRIBUTION SYSTEM DESIGN

This chapter examines different applications of the consumptive demand models specified in Chapter 5 and estimated in Chapter 6. These application areas include:

- Planning Distribution Systems for Growth
- Allocating Demand as an input to Distribution Models
- Reliability – Quantity and Quality

PLANNING DISTRIBUTION SYSTEMS FOR GROWTH

Planning for new growth requires use of design parameters for the demand load of new connections. The formal models developed in this report can be applied to planning new or augmented distribution systems in several informative ways:

1. Duty Factor: forecasting maximal demand load
2. Attaching a probability distribution to demand load
3. Testing for long term changes in demand load

One application of the consumptive demand models is to forecast the distribution of demand load[1]. As seen in the previous chapter, water demand varies significantly by the type of connection (customer class). Even within relatively the most homogeneous class of single-family customers, water demand varies greatly depending on the temperature/weather zone and the connection lot size. The estimated models strongly suggest that the maximum demand load design parameter (duty factor) can be improved by allowing it to (1) vary across customer classes and (2) vary by other significant predictors of water consumption (temperature/weather zone, lot size, number of units or meter size). Since key distribution system design criteria are formed form combining spatial demand load with fire flow requirements, an accurate definition of spatial demand is key to minimizing infrastructure costs.

A second application of the consumptive demand models relates to formally defining likelihood of different demand levels. An estimate of maximum daily demand taken from a short time period of consumption history—one to three years—will only reflect demand under the observed weather in that history. Using the demand models developed in this report, it is possible to backcast demand over a much wider range of climate conditions (50-100 years). This makes possible a more accurate estimate of the probability distribution surrounding demand load and, hence, a more accurate estimate of maximal demand in the tail of the distribution.

A third application of the consumptive demand models involves detecting and modeling changes in long term demand load. Long-term demand models predict changes in water demand based on changes in population, household density, income, and price-induced and non-price induced conservation. The integration of the predictive power of existing long-term planning

---

[1] For a GIS-based approach to demand forecasting, see the Water Demand Forecaster v. 1.3 presentation at http://www.dcse.com.

models offers the potential to improve the definition of demand load in the base file data that drive network system models.

ALLOCATING DEMAND AS AN INPUT TO DISTRIBUTION MODELS

One of the key inputs into modern distribution system models is a valid definition of demand load at each connection/node of the system. These demand data

1. Potential improvements to allocating demand by production interpolation
2. A Database of Demand Load Scenarios for Network Modeling

Improving Demand Allocation

Description of a Typical Demand Allocation Method

A typical method used to translate meter read consumption into a daily demand load estimate uses a measure of daily system production data. In its simplest form, a fraction is formed for each billing period representing the ratio of a day's water production to total production in the period. This ratio then forms the basis for allocating consumption in a billing period to a particular day. A final adjustment includes a system-wide production loss adjustment.

Deficiencies of the Method

1. This method uses one year of consumption history. To the extent that future distribution peaking events exactly replicate the maximum day flow in the selected year, these data will then constitute the correct basis for modeling. To the extent that the modeling is expected to produce information on events not experienced in the selected year, the distribution modeler is at a complete loss. Predictable trends in water consumption will also be left uncaptured.
2. The interpolation method will necessarily embed any errors in production measures in the new demand construct.
3. The use of a spatially undifferentiated estimate of production loss can introduce unnecessary error. An average system loss could be far off for specific subareas of the service area even if it is the exact calibration factor for the service area.
4. This method provides no basis for formal scientific inference. The geographical map of demand load is not attached to any probability distribution. The sampling distribution of consumption data from a single year—given weather-driven variations in demand—will almost certainly underestimate the distribution of consumption taken from a larger period of time. (Strictly speaking, the sample of maximum day peak in a single year contains exactly 1 observation. Inference is quite limited from a sample size of one.)

Suggestions for Improvements

Data manipulation and validation. Attention to the data generation process can correct for known measurement errors. Inclusion of additional logic to account for varying meter read quality will improve the resultant demand estimates from consumption billing history. Estimated reads convey no information on the amount of water use in a billing period. Model based diagnostics of unusual consumption observations can also prevent corrupted data from corrupting network models. Another source of error in consumption records comes from the decreasing precision of water meters as they age. Older water meters tend to under register low levels of flow. Many utilities maintain electronic records of their meter replacement programs, separate from billing system consumption data, that can be used to adjust consumption reads. Meter mismeasurement is, of course, just one component of unaccounted for water. Better tracking of differences between water produced and water sold can lead to an improved understanding of production losses.

Spatially differentiated daily interpolation. Developing additional spatial differentiation for the daily interpolation would also constitute a major improvement. This involves use of subarea daily production measures. The production loss can then be treated as a time series and examined for spatial patterns. This increase in geographical resolution can have material effects on the estimated demand loads used to drive network optimization.

Customer Class-based Characteristic Curves. The existence of different demand load patterns by customer class admits another possible improvement. Applying the customer class characteristic load curves (graphed in the previous chapter) would allow a refined time-allocation of meter read consumption data. During a hot weather, high demand, peaking event it is likely that single family customers will constitute a higher proportion of the demand load in a mixed sub-area containing both single family customers and multifamily customers.

Demand Load Scenarios

Even after all correctable improvements have been made to the current demand allocation/interpolation methods; they should still be viewed as a starting point. Other informative demand scenarios that may be required to develop the most fully informative set of network optimization runs. These additional demand scenarios should (1) be better tied to long run demand forecasts that incorporate the effect on nodal demand of predictable land use changes and water end use changes and (2) have a formal and scientific basis for making inference about the reliability of a prospective distribution system meeting demand. The following section discusses use of more than one demand load scenario, typically peak flow conditions, in the application of water quality modeling.

RELIABILITY – QUANTITY AND QUALITY

Planning for distribution system reliability can be divided into two separate but related topics: planning for delivering a reliable quantity and planning for delivering a reliable level of water quality. Because many of the water quality indicators are defined as a concentration, issues of water quantity always enter into analysis of water quality. The formal models developed in this report can be applied to both of these reliability planning problems.

1. Delivering a Reliable Quantity—Analysis of Outage Management
2. Delivering a Reliable Quality—The Demands of Quality Modeling

Delivering a Reliable Quantity

In the not too distant future, water system planners and managers can expect changes in customer involvement in decisions over how much to invest in delivery reliability. The trend of increasing customer involvement will especially hold true when water systems confront expensive investment decisions in water main replacement or rehabilitation. The need for an empirically sound understanding of customer demand characteristics needs to be integrated into analysis of potential water delivery outages and outage management. Thus, the understanding of demand must extend beyond the observed historical consumption and ask, "What is the additional delivery reliability for which customers are willing to pay?" Some water utilities have conducted "willingness-to-pay" customer surveys to determine the characteristics of water service that customers value. Being able to spatially depict customer valuation of water—that is water demand in the economic sense of the term—constitutes the next step in planning for improved drinking water distribution systems.

There is also exciting progress being made on improved network optimization methods. The use of genetic algorithm methods in network analysis has advanced considerably.[2] These methods are being used in utilities across the globe to minimize replacement and rehabilitation costs. The increased capabilities of network optimization models do not free them from reliance on high quality input data. To the contrary, higher quality baseline demand load files allow these models to use their full capabilities.

Delivering a Reliable Quality

Due to recent changes in drinking water regulations and increased customer interest, water utilities will need to think harder about how to deliver improved water quality. The objective of delivering reliable water quality has increased the need to rehabilitate or replace distribution system components. The increased attention that will be paid to improving delivered water quality has two implications for the need to understand demand load. First, water quality issues mean that *minimum* demand load (the other side of the demand probability distribution) now enters into design criteria for new or rehabilitated distribution systems. Second, the modeling of water quality places greater stress on the definition of water consumption in the base data-file. Each will be discussed in turn.

The fact that the financial stakes are large for sizing and replacement decisions means that the costs for oversizing components are no longer insignificant. Since water quality is affected by the standing time of water in pipes, minimum levels of demand load becomes an important criteria in sizing decisions. (Oversized water mains become, in effect, water storage where excessive water stand time can degrade water quality.) Water utilities that have a good handle on demand load's spatial distribution will be better placed to make wise rehabilitation investment decisions.

Modeling the state of drinking water quality in a distribution system is an even more complex task for network models. Modeling the state of water quality requires knowledge of more than maximal demand load. Incorporating predictions of major nutrients, disinfection

---

[2] See, for example, Frey, McCuller, Henke, Spackman and Murphy (2001) or Zheng , Boulos, Orr, and Ro (1998) for examples of GA network optimization approaches.

byproducts, or microbes can require knowledge of demand load throughout the year. Calibration of network models to a single peak period can provide insufficient accuracy to other nonpeak demand states. It can be expected that water utilities will also find greater payoffs to investments in network optimization models that address both quality and quantity.

CHAPTER 8
FINDINGS

1. Water consumption data, though messy, can provide spatially more precise estimates of demand load for use in distribution system models.

2. Attention to data quality and sample design permit improved inference.

3. Statistical methods and models provide a useful means to

    a. Interpolate average daily consumption through time
    b. Provide a formal basis for probabilistic statements about demand load
    c. Formally test for the existence of spatial differences and trends in consumption
    d. Quantify the effect of weather on demand load (maximal effect within consumption sample and maximal effect throughout recorded weather history.)

4. Additional applications of these methods include:

    a. Incorporating long-term water demand growth drivers into distribution system planning.
    b. Using consumption models to populate demand load scenarios for use in more demanding network optimization models.
    c. Applying consumption models for improved quantity and quality reliability planning for water distribution systems

APPENDIX A
SURVEY OF UTILITY PRACTICES

Part of the project was to conduct a literature search, utility survey and review of existing practices to obtain information on spatial demand analysis methods practiced across the U.S. by water utilities. This appendix presents the results of the utilities survey.

A nonrandom sample of utilities were selected from 893 AWWA utilities to provide information on spatial demand analysis methods as practiced across the U.S. Since this was not a scientific survey, no formal inference will be attempted. Budgetary constraints limited the potential sample size to about 100 utilities. Inference is further limited by a low response rate. Though one cannot draw formal conclusions from this limited sample to the population of all U.S. utilities, it still provides useful insight on prevailing practices.

METHODOLOGY

A subset of 104 utilities were selected from 893 AWWA utilities based on the following characteristics and criteria:

- Geographic Region Within the U.S
- Population Served
- Delivery of Surface Water/Groundwater
- Delivery for Agricultural Use

A survey form with an explanatory cover letter (at the end of this Appendix) was prepared and mailed to the selected group of utilities. Responses were received from utilities that chose to (or were able to) participate. The survey response results were stored in a database for ease of data management, analysis and reporting.

RESULTS AND CONCLUSIONS

Responses

- Only 15 out of 104 utilities returned survey responses. AWWA utilities need improvement in participating in surveys such as these. Educational outreach efforts emphasizing the importance of water-industry research surveys are needed.

Characteristics of the 15 Utilities that Participated

- Of the 15 utilities that participated, utilities from the East (3) and the South (2) were less well represented than the West (5) and Midwest (5) regions.
- Approximately half (7) of the participating utilities deliver only surface water, and about half (8) deliver both surface water and groundwater.
- Utilities that deliver a portion of their supply for agricultural use (4) were underrepresented.

Current Demand Forecast Methodology and Data Management

- **Processes followed by utilities in generating spatial demand forecasts**

  The demand forecasting processes reported vary from:
  - rudimentary methods such as assumed fixed annual percent increase, to
  - intermediate simplicity using some form of demographics and/or moderate system disaggregation, to
  - more sophisticated techniques utilizing detailed geographic sub-areas, demographics, land use, and/or water-use category breakdowns

- **Specific computer applications, computer modeling efforts or statistical analyses used by utilities**

  Few utilities answered this portion of the survey. Mentioned were:
  - current statistical regressions,
  - current hydraulic modeling, and
  - movement toward more GIS and hydraulic model use.

- **Organizational unit within utilities responsible for spatial demand forecasts, and/or consultants utilized by utilities to conduct spatial demand forecasting**

  - It was not always possible to tell, but generally in decreasing order, units answering the survey were engineering, followed closely by planning, then operations or unknown.
  - About half the utilities use consultants in place of or in addition to staff.

- **Demand forecasting categories used by utilities**

  - In decreasing order, the categories of "Aggregate System Demand", "Geographical Region or Subarea" and "Municipal, Industrial, and Agricultural Demand" are the most predominant.

- **Demand forecasting methods used by utilities**

  - In decreasing order, "Per Capita" and "Land-Use Based" are the most predominantly used methods.

- **Demand forecasting period(s) used by utilities**

  - The three most common periods are 5, 10, and 20 years.

- **Computer models/packages utilities use for hydraulic modeling**

  - The two most frequently used models are H2Onet and Cybernet/WaterCAD (WaterCAD is a later-generation evolution of Cybernet, from the same company)

- **Computer models/packages utilities use for spatial demand forecasting**
  - In decreasing order, the most commonly used software are GIS (mostly ESRI products), AutoCAD and Excel.
- **Methods used by utilities to input spatial demand forecasts into hydraulic models**
  - Manual input dominates, with import/export files closely following.
- **Variables taken into account in demand forecasts by utilities**
  - "Population Growth" is by far the most used, with much less usage of "Regional Economic Forces" and "Climate".
- **Basic equations that describe utilities' forecasting methods**

  Few utilities answered this portion of the survey. Equations vary from:
  - simplistic water-use categories or units multiplied by unit factors, to
  - linear interpolation of peak and average day values, to
  - non-linear, empirical parametric formulas.
- **Utilities using or planning to use recycled water in any portions of their service area, and how they account for demands for the recycled water**
  - Only 3 of the 15 utilities that responded use or plan to use recycled water.
  - No clear descriptions of how they account for demands for the recycled water were provided.
- **Annual reports, planning documents, engineering reports or water management reports provided by the utilities**
  - Only three utilities provided documents.

SURVEY SAMPLE DESIGN

The 100 utilities were selected based on the following hierarchy of characteristics and criteria:

Geographic Region Within the U.S.

The geographic regions selected were: East (E), West (W), Midwest (MW), and South (S). About 25 utilities were chosen from each region.

The division of the U.S. into these 4 regions was based on aggregation of existing sub-regions defined by the Federal government:

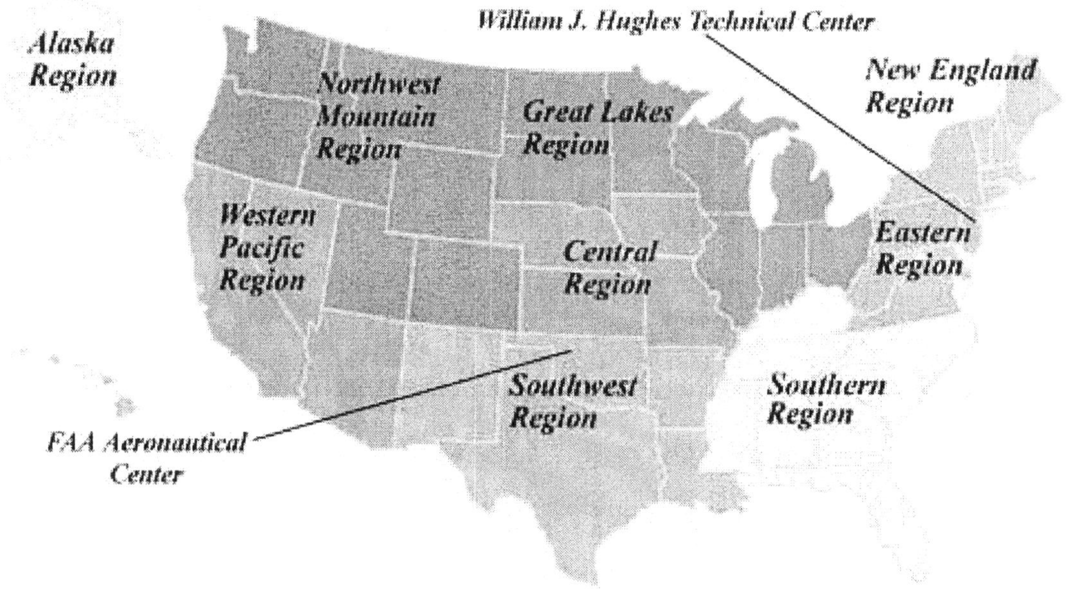

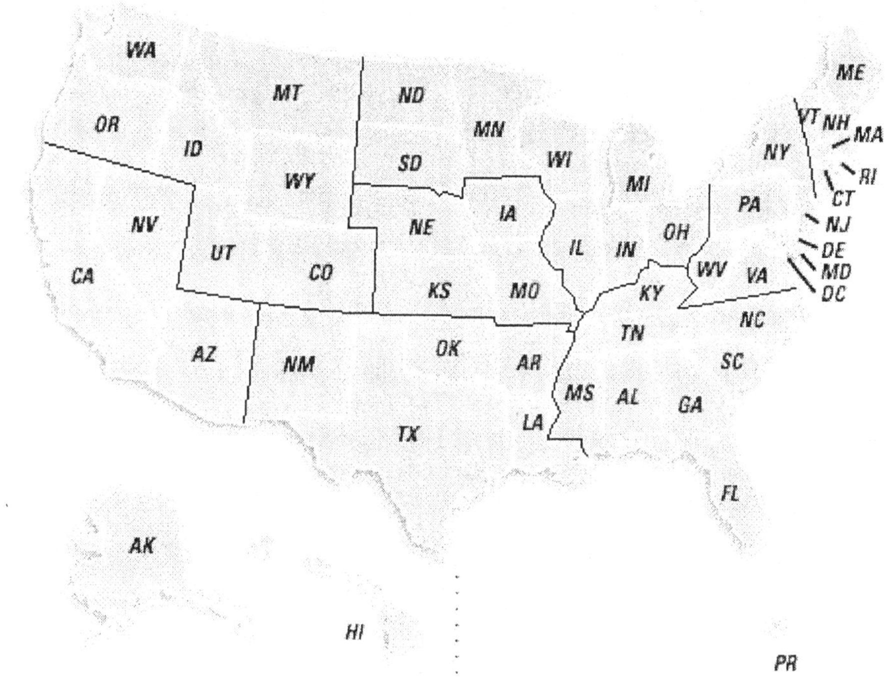

The aggregation of individual states by sub-regions into the 4 general regions was defined as follows:

- East
    - New England Region: Connecticut, Maine, Massachusetts, New Hampshire, Rhode Island and Vermont.

  - Eastern Region: Delaware, Maryland, New Jersey, New York, Pennsylvania, Virginia, West Virginia and the District of Columbia

- West

  - Western Pacific Region: Arizona, California and Nevada
  - Northwest Mountain Region: Colorado, Idaho, Montana, Oregon, Utah, Washington and Wyoming

- Midwest

  - Great Lakes Region: Illinois, Indiana, Michigan, Minnesota, North Dakota, Ohio, South Dakota and Wisconsin
  - Central Region: Iowa, Kansas, Missouri and Nebraska

- South

  - Southern Region: Alabama, Florida, Georgia, Kentucky, Mississippi, North Carolina, South Carolina and Tennessee
  - Southwest Region: Arkansas, Louisiana, New Mexico, Oklahoma and Texas

Population Served

For each set of 25, about 8 each were selected using 3 categories of: Small (S), Medium (M) and Large (L) in terms of the population served.

These 3 categories were determined by ranking the 893 AWWA "WATER\:STATS" utilities in descending order of population served. 26 utilities had no information (number) for population served, so actually only 867 utilities could be used for the ranking. From this ranking, the one-third (33.33%) percentiles were determined. The upper percentile was designated as Large utilities, the middle percentile was designated as Medium, and the lower percentile was designated as Small.

Delivery of Surface Water/Groundwater

For each set of 8, about 4 utilities delivering only surface water (SW), and 4 delivering both surface water and groundwater (SW+GW), were selected.

Delivery for Agricultural Use

For each set of 4, about 2 each that deliver a portion of their supply for agricultural use (Ag), and 2 that do not (Non-Ag), were chosen.

Selection Logistics

A tabulation checklist form that summarizes the selection hierarchy was prepared for each of the four geographic regions, with 25 utilities each:

1	L	SW	Non Ag
2	L	SW	Non Ag
3	L	SW	Non Ag
4	L	SW	Ag
5	L	SW	Ag
6	L	SW+GW	Non Ag
7	L	SW+GW	Non Ag
8	L	SW+GW	Ag
9	L	SW+GW	Ag
10	M	SW	Non Ag
11	M	SW	Non Ag
12	M	SW	Ag
13	M	SW	Ag
14	M	SW+GW	Non Ag
15	M	SW+GW	Non Ag
16	M	SW+GW	Ag
17	M	SW+GW	Ag
18	S	SW	Non Ag
19	S	SW	Non Ag
20	S	SW	Ag
21	S	SW	Ag
22	S	SW+GW	Non Ag
23	S	SW+GW	Non Ag
24	S	SW+GW	Ag
25	S	SW+GW	Ag

To the extent possible, this breakdown of selected utilities by characteristics was followed. However, some geographic regions did not have 25 utilities that matched these combined criteria. Specifically, for some regions, there were not enough surface water/groundwater-supplying and/or agricultural-supplying utilities to exactly match the statistical-design checklist. In those cases, substitutions had to be made (to maintain 25 per region) with other utilities to match the design as closely as possible.

The results of the initial selection of 100 utilities and their characteristics, separated into the 4 geographic regions, was as follows:

Utility Name	Region	Size	SW/SW+GW	Ag/Non-Ag
Erie County Water Auth	E	L	SW	Non-Ag
Hartford Metropolitan Dist	E	L	SW	Non-Ag
Washington Aqueduct	E	L	SW	Non-Ag
Washington Suburban Sanit Comm	E	L	SW	Non-Ag
New Bedford, (City of) Water Dept	E	L	SW	Ag
Philadelphia Suburban Water Co	E	L	SW+GW	Non-Ag
Elizabethtown Water Co	E	L	SW+GW	Non-Ag
Fairfax County Water Auth	E	L	SW+GW	Non-Ag
Brick Township Muni Util Auth	E	L	SW+GW	Ag
Bangor Water Dist	E	M	SW	Non-Ag
Champlain Water Dist	E	M	SW	Non-Ag
Coatesville, (City of) Auth	E	M	SW	Ag
Haverhill Water Dept	E	M	SW	Ag
Danbury, (City of)	E	M	SW+GW	Non-Ag
North Penn Water Auth	E	M	SW+GW	Non-Ag
Consumers PA Water Co	E	M	SW+GW	Non-Ag
Abington-Rockland Joint WW	E	M	SW+GW	Ag
Harrison Township Water Auth	E	S	SW	Non-Ag
Kingston Water Dept	E	S	SW	Non-Ag
Kittery Water Dist	E	S	SW	Ag
Oneida, (City of) Water Dept	E	S	SW	Ag
Keene, (City of) Water Dept	E	S	SW+GW	Non-Ag
Augusta Water Dist	E	S	SW+GW	Non-Ag
Bristol County Water Auth	E	S	SW+GW	Non-Ag
Citizens Util Water Co of PA	E	S	SW+GW	Non-Ag
Cleveland Div of Water	MW	L	SW	Non-Ag
St. Louis County Water Co	MW	L	SW	Non-Ag
Chicago, (City of) Dept of Water	MW	L	SW	Non-Ag
Milwaukee Water Works	MW	L	SW	Non-Ag
Kenosha Water Util	MW	L	SW	Ag
Kansas City Water Svcs Dept	MW	L	SW+GW	Non-Ag
Omaha Metro Util Dist	MW	L	SW+GW	Non-Ag
Indianapolis Water Co	MW	L	SW+GW	Non-Ag
Columbus Water Div	MW	L	SW+GW	Non-Ag
Minot Water Dept	MW	M	SW	Non-Ag
Highland Park, (City of) Water	MW	M	SW	Non-Ag
Council Bluffs Water Works	MW	M	SW	Non-Ag
Saint Cloud Public Util	MW	M	SW	Non-Ag
Leavenworth Water Dept	MW	M	SW+GW	Non-Ag
Iowa City Water Department	MW	M	SW+GW	Non-Ag
Consumers Illinois Water Co	MW	M	SW+GW	Non-Ag
IN-American Water Co Inc	MW	M	SW+GW	Non-Ag
Oak Creek Water & Sewer Auth	MW	S	SW	Non-Ag
Neehah City Water Works	MW	S	SW	Non-Ag
Aberdeen, (City of)	MW	S	SW	Non-Ag
Oregon, (City of) Water Plant	MW	S	SW	Non-Ag
Fort Madison Water Dept	MW	S	SW+GW	Non-Ag
Neosho, (City of)	MW	S	SW+GW	Non-Ag
Logansport Muni Util	MW	S	SW+GW	Non-Ag
Valparaiso Dept Water Works	MW	S	SW+GW	Non-Ag

Charlotte-Mecklenburg Util	S	L	SW	Non-Ag
Tulsa Public Works Dept	S	L	SW	Non-Ag
DeKalb County Public Works	S	L	SW	Non-Ag
Louisville Water Company	S	L	SW	Non-Ag
New Orleans Water/Sewer Board	S	L	SW	Non-Ag
El Paso Water Util	S	L	SW+GW	Non-Ag
Tampa Water Dept	S	L	SW+GW	Non-Ag
Houston, (City of)	S	L	SW+GW	Non-Ag
Amarillo Muni Water System	S	L	SW+GW	Ag
Fort Myers, (City of)	S	M		Ag
Lenoir, (City of)	S	M	SW	Non-Ag
Columbia Water System	S	M	SW	Non-Ag
St. Bernard Parish	S	M	SW	Non-Ag
Wilmington Water & Sewer	S	M	SW	Ag
Midwest City Water Dept	S	M	SW+GW	Non-Ag
Johnson City Water Dept	S	M	SW+GW	Non-Ag
Nacogdoches Water Dept	S	M	SW+GW	Non-Ag
Greenville Water Dept	S	S	SW	Non-Ag
Morganton, (City of)	S	S	SW	Non-Ag
Punta Gorda Water Treatment	S	S	SW	Ag
Cash Water Supply Inc	S	S	SW	Ag
Oneonta Util Board	S	S	SW+GW	Non-Ag
Hope Water & Light Comm	S	S	SW+GW	Non-Ag
Okeechobee Util Auth	S	S	SW+GW	Non-Ag
Fayetteville Water & Sewer	S	S	SW+GW	Non-Ag
Eugene Water & Electric Board	W	L	SW	Non-Ag
Tempe, (City of) Water Mngmt Div	W	L	SW	Non-Ag
Fort Collins Water Util	W	L	SW	Non-Ag
Contra Costa Water Dist	W	L	SW	Ag
Pueblo Board of Waterworks	W	L	SW	Ag
Salt Lake City Public Util	W	L	SW+GW	Non-Ag
Seattle Water Dept	W	L	SW+GW	Non-Ag
Corona, (City of) Util Svcs	W	L	SW+GW	Ag
CA-American Water Co	W	L	SW+GW	Ag
Yuma, (City of) Public Works Dept	W	M	SW	Non-Ag
Great Falls, (City of)	W	M	SW	Non-Ag
Bellingham, (City of)	W	M	SW	Ag
Public Util Dist #1 Skagit Co	W	M	SW	Ag
Centennial Water & Sanit Dist	W	M	SW+GW	Non-Ag
Casper, (City of)	W	M	SW+GW	Non-Ag
Rialto Water Dept	W	M	SW+GW	Ag
Casitas Muni Water Dist	W	M	SW+GW	Ag
Canon City Water Dept	W	S	SW	Non-Ag
Evanston Water Dept	W	S	SW	Non-Ag
Santa Fe Irrigation Dist	W	S	SW	Ag
Valley Center Water Dist	W	S	SW	Ag
Mammoth Community Water Dist	W	S	SW+GW	Non-Ag
Lewiston, (City of) Water Dept	W	S	SW+GW	Non-Ag
Carpinteria Co Water Dist	W	S	SW+GW	Ag
East Niles Comm Svcs Dist	W	S	SW+GW	Ag

COVER LETTER AND SURVEY FORM

Mr/s. Xxxxxxxx
Xxxxx Water District
Xxxx Xxxx Street
City, State Zip

Date

RE: AWWARF Spatial Demand Allocation Research Project

Dear Mr/s. Xxxxxxxx

The American Water Works Association Research Foundation (AWWARF) is sponsoring a tailored collaboration research to address problems associated with identification and spatial allocation of water demands used for distribution system planning and design. The collaborative research effort is managed by the Los Angeles Department of Water and Power (LADWP). The other members of the research team are: Metropolitan Water District of Southern California (Metropolitan), Irvine Ranch Water District (IRWD), Calleguas Municipal Water District (Calleguas), A&N Technical Services, Inc. (A&N), and Diba Consulting Software Engineers (DCSE).

DCSE is conducting a literature search, utility survey and review of the existing practices to obtain information on spatial demand analysis methods practiced throughout the region. The project team has prepared a survey form and has mailed the form to a selected group of agencies. Your agency is one of the selected agencies. The project team would greatly appreciate your assistance in completing the attached survey form and returning it to us.

Demand allocation constitutes a critical component of the water distribution design and planning process. Your input is very valuable and we are confident it will contribute to the success of this research study. Please feel free to contact me at (949) 347-7584 or msh@dcse.com should you have any questions or need more information.

Sincerely,

Masoud S. Hoseyni, Ph.D.
Vice President
DCSE

SPATIAL DEMAND ALLOCATION FOR DISTRIBUTION SYSTEM DESIGN SURVEY

AGENCY: ________________________________

DATE: ________________________________

<u>Current Demand Forecast Methodology and Data Management</u>

1) Describe the process followed by your agency in generating spatial demand forecast. Does this process include any specific computer applications, computer modeling efforts or statistical analyses?

____

____

____

____

____

____

____

____

____

____

____

____

____

____

Which organizational unit within your utility performs spatial demand forecasts?

____

Contact name_______________________ Phone no. ________________

Or does your utility utilize consultants to conduct spatial demand forecasting?
Consulting firm__________________________________
Contact name_______________________ Phone no. ________________

Demand Forecasting Categories

2) Aggregate System Demand
3) Urban and Agricultural Demand
4) Municipal, Industrial, and Agricultural Demand
5) Geographical region or Subarea
6) Other (Please specify) ____________________

Demand Forecasting Methods

7) Per Capita
8) Land Use Based
9) Econometrics/Demographics
10) Other (Please specify) ____________________

11) Demand Forecasting Period(s)

12) One (1) Year
13) Five (5) Years
14) Ten (10) Years
15) Twenty (20) Years
16) Thirty (30) Years
17) Other (Please specify) ____________________

18) Please list/describe the computer models/packages you use for Hydraulic modeling (e.g., WaterCAD, H_2ONet,...)

____

____

____

19) Please list/describe the computer models/packages you use for spatial demand forecasting (e.g., GIS, Statistical models...)

____

____

20) Please describe how do you input your spatial demand forecasts into the Hydraulic model? (Manual input, import/export files, computer interface)

21) Do your forecasts account for variables such as (*please check the ones that apply to your agency*):

22) Population growth
23) Climate
24) Regional economic forces
25) Personal income
26) Price of water

Please provide a brief description:

27) If possible, please write the basic equations that describe your forecasting method

____

____

____

28) Describe your agency's typical system operating plan (i.e. do you purchase imported water, how are local resources used in conjunction with imported resources, etc.).

____

____

____

____

29) Do you use or planning to use recycled water in any portions of your service area? If yes, please indicate how do you account for demands for the recycled water?

____

____

____

____

30) Please provide copies of reports, annual reports, planning documents, engineering reports, or water management reports for your agency.

APPENDIX B
SINGLE FAMILY CONSUMPTION MODELS BY SUBAREA

This appendix contains modes for single-family consumption broken out by hydraulic submodel keyed to pressure service zones.

Table B.1

Submodels and Pressure Service Zones

SubArea	SUBMODEL	PRESSURE SERVICE ZONES
1	Foothills	1010, 1051, 1275, 1449, 1500, 1720, 2045.
2	Sunland/Tujunga	1597, 1670, 1749, 1960, 1990, 2086, 2140, 2200, 2440
3	Granada Hills	1750, 1615, 1445, 1200, 1305, 1410, 1383, 1540, 1383, 1540
4	Valley Floor	947, 1075, 1123, 1134, 1000, 830
5	Encino Hills	1240, 1337, 1677
6	Santa Monica Mountains	1636, 1350, 1216, 1424, 1096, 1546, 1600, 1050, 801, 825
7	Hollywood Hills	1116, 757, 1232, 1280, 1143, 946, 1030, 765, 721
8	Westside	205, 295, 325, 420, 440, 426, 540, 610, 740, 847, 865, 930.
9	Santa Ynez	310, 330, 375, 498, 529, 720, 1137, 1345, 1645
10	Mount Washington	619, 769, 675, 808, 870, 944, 1070, 667, 752, 730, 517
11	Highland Park	950, 778, 1066, 890, 801, 737
12	Central City	275, 386, 462, 477, 448, 579, 585, 670
13	Harbor	240, 320, 365, 400, 597, 655,770

Table B.2

Single Family Residential Water Consumption Model
SubArea 1—Foothills

Dependent Variable: Log Average Daily Metered Water Consumption
(in gallons per day)

Independent Variable	Coefficient	Std. Error
1. First Sine harmonic, 132 month (annual) frequency	-0.19070	0.00920
2. First Cosine harmonic, 12 month (annual) frequency	-0.32887	0.00948
3. Second Sine harmonic, 6 month frequency	0.00123	0.01071
4. Second Cosine harmonic, 6 month frequency	-0.01501	0.01095
5. Third Sine harmonic, 4 month frequency	0.00379	0.01667
6. Third Cosine 4month frequency	0.00385	0.01031
7. Fourth Sine harmonic, 3 month frequency	0.04506	0.02137
8. Fourth Cosine, 3 month frequency	0.02754	0.02118
9. Fifth Sine harmonic	0.05228	0.04525
10. Fifth Cosine harmonic	0.01644	0.04380
11. Interaction of contemporaneous temperature with annual sine harmonic	0.43842	0.16706
12. Interaction of contemporaneous temperature with annual cosine harmonic	-0.21667	0.21529
13. Deviation from ln(31 or 61 day moving average of maximum daily air temperature)	1.10540	0.12267
14. Interaction of contemporaneous rain with annual sine harmonic	-0.03934	0.01192
15. Interaction of contemporaneous rain with annual cosine harmonic	-0.04488	0.01562
16. Deviation from ln(31 or 61 day moving sum of rainfall+1)	-0.02909	0.01009
17. Bi-monthly lag from rain deviation	-0.01997	0.00606
18. Constant (Mean intercept)	5.84854	0.00376
Standard Error of Individual Constant Terms	0.57927	
Standard Error of White Noise Error	0.32382	
Number of observations	9223	
Number of customer accounts	322	
Time period	**January 1992- August 2000**	

Table B.3
Single Family Residential Water Consumption Model
SubArea 2—Sunland/Tujunga

Dependent Variable: Log Average Daily Metered Water Consumption (in gallons per day)

Independent Variable	Coefficient	Std. Error
1. First Sine harmonic, 132 month (annual) frequency	-0.17080	0.01886
2. First Cosine harmonic, 12 month (annual) frequency	-0.33915	0.01816
3. Second Sine harmonic, 6 month frequency	-0.00203	0.02685
4. Second Cosine harmonic, 6 month frequency	0.00406	0.02644
5. Third Sine harmonic, 4 month frequency	0.01484	0.05084
6. Third Cosine 4month frequency	0.02233	0.01067
7. Fourth Sine harmonic, 3 month frequency	0.00699	0.05469
8. Fourth Cosine, 3 month frequency	-0.01428	0.05082
9. Fifth Sine harmonic	0.15071	0.08724
10. Fifth Cosine harmonic	0.11716	0.08347
11. Interaction of contemporaneous temperature with annual sine harmonic	0.47408	0.25325
12. Interaction of contemporaneous temperature with annual cosine harmonic	-0.08206	0.37624
13. Deviation from ln(31 or 61 day moving average of maximum daily air temperature)	1.10749	0.19629
14. Interaction of contemporaneous rain with annual sine harmonic	-0.04733	0.02265
15. Interaction of contemporaneous rain with annual cosine harmonic	-0.07722	0.02616
16. Deviation from ln(31 or 61 day moving sum of rainfall+1)	-0.02322	0.01916
17. Bi-monthly lag from rain deviation	-0.03696	0.01012
18. Constant (Mean intercept)	5.69991	0.00630
Standard Error of Individual Constant Terms	0.75997	
Standard Error of White Noise Error	0.35088	
Number of observations	4218	
Number of customer accounts	154	
Time period	**January 1992- August 2000**	

Table B.4

Single Family Residential Water Consumption Model
SubArea 3—Granada Hills
Dependent Variable: Log Average Daily Metered Water Consumption
(in gallons per day)

Independent Variable	Coefficient	Std. Error
1. First Sine harmonic, 132 month (annual) frequency	-0.23488	0.00532
2. First Cosine harmonic, 12 month (annual) frequency	-0.36050	0.00516
3. Second Sine harmonic, 6 month frequency	-0.03419	0.00580
4. Second Cosine harmonic, 6 month frequency	0.00840	0.00540
5. Third Sine harmonic, 4 month frequency	-0.01643	0.00580
6. Third Cosine 4month frequency	0.01867	0.00804
7. Fourth Sine harmonic, 3 month frequency	0.03298	0.01077
8. Fourth Cosine, 3 month frequency	0.02249	0.01075
9. Fifth Sine harmonic	0.01366	0.02292
10. Fifth Cosine harmonic	-0.10527	0.02262
11. Interaction of contemporaneous temperature with annual sine harmonic	0.63415	0.12977
12. Interaction of contemporaneous temperature with annual cosine harmonic	0.30556	0.16552
13. Deviation from ln(31 or 61 day moving average of maximum daily air temperature)	1.23343	0.09853
14. Interaction of contemporaneous rain with annual sine harmonic	-0.02732	0.01092
15. Interaction of contemporaneous rain with annual cosine harmonic	-0.01506	0.01379
16. Deviation from ln(31 or 61 day moving sum of rainfall+1)	-0.09397	0.00955
17. Bi-monthly lag from rain deviation	-0.02729	0.00556
18. Constant (Mean intercept)	6.15399	0.00325
Standard Error of Individual Constant Terms	0.72043	
Standard Error of White Noise Error	0.32314	
Number of observations	13410	
Number of customer accounts	473	
Time period	January 1992- August 2000	

Table B.5

Single Family Residential Water Consumption Model
SubArea 4—Valley Floor

Dependent Variable: Log Average Daily Metered Water Consumption (in gallons per day)

Independent Variable	Coefficient	Std. Error
1. First Sine harmonic, 132 month (annual) frequency	-0.19481	0.00209
2. First Cosine harmonic, 12 month (annual) frequency	-0.32905	0.00203
3. Second Sine harmonic, 6 month frequency	-0.01190	0.00228
4. Second Cosine harmonic, 6 month frequency	0.00557	0.00212
5. Third Sine harmonic, 4 month frequency	-0.01062	0.00257
6. Third Cosine 4month frequency	0.01701	0.00287
7. Fourth Sine harmonic, 3 month frequency	0.01394	0.00418
8. Fourth Cosine, 3 month frequency	0.00460	0.00418
9. Fifth Sine harmonic	0.01846	0.00889
10. Fifth Cosine harmonic	-0.06347	0.00888
11. Interaction of contemporaneous temperature with annual sine harmonic	0.53574	0.05463
12. Interaction of contemporaneous temperature with annual cosine harmonic	-0.23237	0.06983
13. Deviation from ln(31 or 61 day moving average of maximum daily air temperature)	0.82273	0.04150
14. Interaction of contemporaneous rain with annual sine harmonic	-0.03814	0.00466
15. Interaction of contemporaneous rain with annual cosine harmonic	-0.03007	0.00572
16. Deviation from ln(31 or 61 day moving sum of rainfall+1)	-0.07519	0.00396
17. Bi-monthly lag from rain deviation	-0.02937	0.00230
18. Constant (Mean intercept)	5.86745	0.00135
Standard Error of Individual Constant Terms	0.57941	
Standard Error of White Noise Error	0.3202	
Number of observations	72925	
Number of customer accounts	2757	
Time period	**January 1992- August 2000**	

Table B.6

Single Family Residential Water Consumption Model
SubArea 5—Encino Hills

Dependent Variable: Log Average Daily Metered Water Consumption (in gallons per day)

Independent Variable	Coefficient	Std. Error
1. First Sine harmonic, 132 month (annual) frequency	-0.25091	0.00750
2. First Cosine harmonic, 12 month (annual) frequency	-0.37331	0.00747
3. Second Sine harmonic, 6 month frequency	-0.03514	0.00830
4. Second Cosine harmonic, 6 month frequency	-0.02243	0.00796
5. Third Sine harmonic, 4 month frequency	-0.01003	0.01173
6. Third Cosine 4month frequency	0.00847	0.00844
7. Fourth Sine harmonic, 3 month frequency	0.02003	0.01563
8. Fourth Cosine, 3 month frequency	0.02225	0.01581
9. Fifth Sine harmonic	-0.15697	0.03277
10. Fifth Cosine harmonic	-0.00508	0.03508
11. Interaction of contemporaneous temperature with annual sine harmonic	0.62457	0.18715
12. Interaction of contemporaneous temperature with annual cosine harmonic	-0.01486	0.24294
13. Deviation from ln(31 or 61 day moving average of maximum daily air temperature)	1.21744	0.14021
14. Interaction of contemporaneous rain with annual sine harmonic	-0.04862	0.01400
15. Interaction of contemporaneous rain with annual cosine harmonic	-0.04031	0.01849
16. Deviation from ln(31 or 61 day moving sum of rainfall+1)	-0.05929	0.01206
17. Bi-monthly lag from rain deviation	-0.03745	0.00715
18. Constant (Mean intercept)	6.30205	0.00441
Standard Error of Individual Constant Terms	0.69323	
Standard Error of White Noise Error	0.3306	
Number of observations	7047	
Number of customer accounts	244	
Time period	**January 1992- August 2000**	

Table B.7
Single Family Residential Water Consumption Model
SubArea 6—Santa Monica Mountains
Dependent Variable: Log Average Daily Metered Water Consumption
(in gallons per day)

Independent Variable	Coefficient	Std. Error
1. First Sine harmonic, 132 month (annual) frequency	-0.18011	0.00606
2. First Cosine harmonic, 12 month (annual) frequency	-0.26118	0.00595
3. Second Sine harmonic, 6 month frequency	-0.01958	0.00703
4. Second Cosine harmonic, 6 month frequency	-0.01209	0.00676
5. Third Sine harmonic, 4 month frequency	-0.00739	0.00918
6. Third Cosine 4month frequency	0.01598	0.00821
7. Fourth Sine harmonic, 3 month frequency	0.02900	0.01345
8. Fourth Cosine, 3 month frequency	0.01196	0.01338
9. Fifth Sine harmonic	-0.05756	0.03221
10. Fifth Cosine harmonic	-0.04463	0.03063
11. Interaction of contemporaneous temperature with annual sine harmonic	0.26610	0.17430
12. Interaction of contemporaneous temperature with annual cosine harmonic	-0.26273	0.21367
13. Deviation from ln(31 or 61 day moving average of maximum daily air temperature)	0.79602	0.13553
14. Interaction of contemporaneous rain with annual sine harmonic	-0.07600	0.01316
15. Interaction of contemporaneous rain with annual cosine harmonic	-0.05249	0.01704
16. Deviation from ln(31 or 61 day moving sum of rainfall+1)	-0.07640	0.01203
17. Bi-monthly lag from rain deviation	-0.04034	0.00713
18. Constant (Mean intercept)	6.24282	0.00380
Standard Error of Individual Constant Terms	0.88713	
Standard Error of White Noise Error	0.33875	
Number of observations	8848	
Number of customer accounts	382	
Time period	**January 1992- August 2000**	

Table B.8
Single Family Residential Water Consumption Model
SubArea 7—Hollywood Hills

Dependent Variable: Log Average Daily Metered Water Consumption (in gallons per day)

Independent Variable	Coefficient	Std. Error
1. First Sine harmonic, 132 month (annual) frequency	-0.15950	0.01180
2. First Cosine harmonic, 12 month (annual) frequency	-0.25410	0.01242
3. Second Sine harmonic, 6 month frequency	-0.01395	0.01303
4. Second Cosine harmonic, 6 month frequency	0.00591	0.01288
5. Third Sine harmonic, 4 month frequency	-0.02934	0.02049
6. Third Cosine 4month frequency	0.03210	0.01044
7. Fourth Sine harmonic, 3 month frequency	0.02775	0.02510
8. Fourth Cosine, 3 month frequency	-0.03589	0.02500
9. Fifth Sine harmonic	0.06843	0.05509
10. Fifth Cosine harmonic	0.02489	0.05347
11. Interaction of contemporaneous temperature with annual sine harmonic	0.36505	0.27183
12. Interaction of contemporaneous temperature with annual cosine harmonic	-0.28006	0.34604
13. Deviation from ln(31 or 61 day moving average of maximum daily air temperature)	0.25660	0.20422
14. Interaction of contemporaneous rain with annual sine harmonic	-0.05055	0.02124
15. Interaction of contemporaneous rain with annual cosine harmonic	-0.08575	0.02531
16. Deviation from ln(31 or 61 day moving sum of rainfall+1)	-0.08328	0.01858
17. Bi-monthly lag from rain deviation	-0.04147	0.01088
18. Constant (Mean intercept)	5.85646	0.00639
Standard Error of Individual Constant Terms	0.72049	
Standard Error of White Noise Error	0.34186	
Number of observations	4130	
Number of customer accounts	144	
Time period	**January 1992- August 2000**	

Table B.9
Single Family Residential Water Consumption Model
SubArea 8--Westside

Dependent Variable: Log Average Daily Metered Water Consumption (in gallons per day)

Independent Variable	Coefficient	Std. Error
1. First Sine harmonic, 132 month (annual) frequency	-0.14757	0.00348
2. First Cosine harmonic, 12 month (annual) frequency	-0.28150	0.00353
3. Second Sine harmonic, 6 month frequency	-0.02849	0.00362
4. Second Cosine harmonic, 6 month frequency	-0.00565	0.00348
5. Third Sine harmonic, 4 month frequency	-0.01236	0.00505
6. Third Cosine 4month frequency	0.01567	0.00393
7. Fourth Sine harmonic, 3 month frequency	-0.00174	0.00704
8. Fourth Cosine, 3 month frequency	-0.00624	0.00692
9. Fifth Sine harmonic	-0.06599	0.01492
10. Fifth Cosine harmonic	-0.07573	0.01509
11. Interaction of contemporaneous temperature with annual sine harmonic	0.02887	0.11691
12. Interaction of contemporaneous temperature with annual cosine harmonic	-0.03693	0.14030
13. Deviation from ln(31 or 61 day moving average of maximum daily air temperature)	0.33451	0.08531
14. Interaction of contemporaneous rain with annual sine harmonic	-0.07136	0.00731
15. Interaction of contemporaneous rain with annual cosine harmonic	-0.01174	0.00958
16. Deviation from ln(31 or 61 day moving sum of rainfall+1)	-0.09336	0.00717
17. Bi-monthly lag from rain deviation	-0.04435	0.00397
18. Constant (Mean intercept)	5.76160	0.00230
Standard Error of Individual Constant Terms	0.71769	
Standard Error of White Noise Error	0.3054	
Number of observations	25668	
Number of customer accounts	981	
Time period	**January 1992- August 2000**	

Table B.10
Single Family Residential Water Consumption Model
SubArea 9--Santa Ynez

Dependent Variable: Log Average Daily Metered Water Consumption (in gallons per day)

Independent Variable	Coefficient	Std. Error
1. First Sine harmonic, 132 month (annual) frequency	-0.02465	0.03716
2. First Cosine harmonic, 12 month (annual) frequency	-0.26187	0.03348
3. Second Sine harmonic, 6 month frequency	0.08094	0.04640
4. Second Cosine harmonic, 6 month frequency	-0.11627	0.04579
5. Third Sine harmonic, 4 month frequency	-0.02905	0.01243
6. Third Cosine 4month frequency	-0.01646	0.08391
7. Fourth Sine harmonic, 3 month frequency	-0.21329	0.09455
8. Fourth Cosine, 3 month frequency	-0.19131	0.09130
9. Fifth Sine harmonic	-0.67878	0.18272
10. Fifth Cosine harmonic	-0.00913	0.15927
11. Interaction of contemporaneous temperature with annual sine harmonic	-0.24322	0.34487
12. Interaction of contemporaneous temperature with annual cosine harmonic	0.88770	0.43975
13. Deviation from ln(31 or 61 day moving average of maximum daily air temperature)	0.95536	0.27341
14. Interaction of contemporaneous rain with annual sine harmonic	-0.04122	0.02374
15. Interaction of contemporaneous rain with annual cosine harmonic	-0.03012	0.02637
16. Deviation from ln(31 or 61 day moving sum of rainfall+1)	-0.15503	0.02099
17. Bi-monthly lag from rain deviation	-0.05736	0.01275
18. Constant (Mean intercept)	6.04014	0.00841
Standard Error of Individual Constant Terms	0.75259	
Standard Error of White Noise Error	0.28758	
Number of observations	3067	
Number of customer accounts	103	
Time period	**January 1992- August 2000**	

Table B.11
Single Family Residential Water Consumption Model
SubArea 10--Mount Washington

Dependent Variable: Log Average Daily Metered Water Consumption (in gallons per day)

Independent Variable	Coefficient	Std. Error
1. First Sine harmonic, 132 month (annual) frequency	-0.09251	0.00619
2. First Cosine harmonic, 12 month (annual) frequency	-0.18445	0.00636
3. Second Sine harmonic, 6 month frequency	0.00714	0.00636
4. Second Cosine harmonic, 6 month frequency	0.00399	0.00620
5. Third Sine harmonic, 4 month frequency	0.00025	0.00723
6. Third Cosine 4month frequency	0.01011	0.00834
7. Fourth Sine harmonic, 3 month frequency	0.00491	0.01195
8. Fourth Cosine, 3 month frequency	0.00540	0.01198
9. Fifth Sine harmonic	-0.05896	0.02550
10. Fifth Cosine harmonic	-0.11627	0.02604
11. Interaction of contemporaneous temperature with annual sine harmonic	-0.12277	0.17041
12. Interaction of contemporaneous temperature with annual cosine harmonic	-0.16973	0.20628
13. Deviation from ln(31 or 61 day moving average of maximum daily air temperature)	-0.11053	0.12964
14. Interaction of contemporaneous rain with annual sine harmonic	-0.02853	0.01415
15. Interaction of contemporaneous rain with annual cosine harmonic	-0.00355	0.01582
16. Deviation from ln(31 or 61 day moving sum of rainfall+1)	-0.07479	0.01190
17. Bi-monthly lag from rain deviation	-0.00990	0.00739
18. Constant (Mean intercept)	5.41120	0.00415
Standard Error of Individual Constant Terms	0.72021	
Standard Error of White Noise Error	0.31957	
Number of observations	9183	
Number of customer accounts	336	
Time period	**January 1992- August 2000**	

Table B.12
Single Family Residential Water Consumption Model
SubArea 11--Highland Park

Dependent Variable: Log Average Daily Metered Water Consumption (in gallons per day)

Independent Variable	Coefficient	Std. Error
1. First Sine harmonic, 132 month (annual) frequency	-0.11511	0.00684
2. First Cosine harmonic, 12 month (annual) frequency	-0.20639	0.00715
3. Second Sine harmonic, 6 month frequency	0.00985	0.00724
4. Second Cosine harmonic, 6 month frequency	0.01639	0.00722
5. Third Sine harmonic, 4 month frequency	-0.00004	0.01098
6. Third Cosine 4month frequency	0.01397	0.00659
7. Fourth Sine harmonic, 3 month frequency	0.00770	0.01387
8. Fourth Cosine, 3 month frequency	-0.01131	0.01384
9. Fifth Sine harmonic	0.00550	0.02860
10. Fifth Cosine harmonic	-0.05078	0.02918
11. Interaction of contemporaneous temperature with annual sine harmonic	0.33394	0.17796
12. Interaction of contemporaneous temperature with annual cosine harmonic	-0.38941	0.22457
13. Deviation from ln(31 or 61 day moving average of maximum daily air temperature)	0.02974	0.13332
14. Interaction of contemporaneous rain with annual sine harmonic	-0.01305	0.01402
15. Interaction of contemporaneous rain with annual cosine harmonic	-0.00897	0.01615
16. Deviation from ln(31 or 61 day moving sum of rainfall+1)	-0.07916	0.01216
17. Bi-monthly lag from rain deviation	-0.02094	0.00703
18. Constant (Mean intercept)	5.40857	0.00419
Standard Error of Individual Constant Terms	0.63499	
Standard Error of White Noise Error	0.32035	
Number of observations	8610	
Number of customer accounts	288	
Time period	**January 1992- August 2000**	

Table B.13
Single Family Residential Water Consumption Model
SubArea 12--Central City

Dependent Variable: Log Average Daily Metered Water Consumption (in gallons per day)

Independent Variable	Coefficient	Std. Error
1. First Sine harmonic, 132 month (annual) frequency	-0.11905	0.00331
2. First Cosine harmonic, 12 month (annual) frequency	-0.22849	0.00339
3. Second Sine harmonic, 6 month frequency	-0.00574	0.00337
4. Second Cosine harmonic, 6 month frequency	0.01010	0.00339
5. Third Sine harmonic, 4 month frequency	-0.00263	0.00437
6. Third Cosine 4month frequency	0.00696	0.00400
7. Fourth Sine harmonic, 3 month frequency	0.01579	0.00639
8. Fourth Cosine, 3 month frequency	-0.00198	0.00642
9. Fifth Sine harmonic	0.01716	0.01350
10. Fifth Cosine harmonic	-0.06887	0.01359
11. Interaction of contemporaneous temperature with annual sine harmonic	0.28844	0.09436
12. Interaction of contemporaneous temperature with annual cosine harmonic	-0.06375	0.11553
13. Deviation from ln(31 or 61 day moving average of maximum daily air temperature)	0.38351	0.07096
14. Interaction of contemporaneous rain with annual sine harmonic	-0.02528	0.00736
15. Interaction of contemporaneous rain with annual cosine harmonic	0.01284	0.00854
16. Deviation from ln(31 or 61 day moving sum of rainfall+1)	-0.06790	0.00636
17. Bi-monthly lag from rain deviation	-0.02267	0.00372
18. Constant (Mean intercept)	5.52349	0.00219
Standard Error of Individual Constant Terms	0.67438	
Standard Error of White Noise Error	0.33587	
Number of observations	34671	
Number of customer accounts	1269	
Time period	**January 1992- August 2000**	

Table B.14
Single Family Residential Water Consumption Model
SubArea 13--Harbor

Dependent Variable: Log Average Daily Metered Water Consumption (in gallons per day)

Independent Variable	Coefficient	Std. Error
1. First Sine harmonic, 132 month (annual) frequency	-0.10763	0.00572
2. First Cosine harmonic, 12 month (annual) frequency	-0.25821	0.00598
3. Second Sine harmonic, 6 month frequency	0.00298	0.00626
4. Second Cosine harmonic, 6 month frequency	0.00547	0.00605
5. Third Sine harmonic, 4 month frequency	0.00666	0.00785
6. Third Cosine 4month frequency	0.00640	0.00752
7. Fourth Sine harmonic, 3 month frequency	0.02035	0.01176
8. Fourth Cosine, 3 month frequency	0.00073	0.01170
9. Fifth Sine harmonic	0.01045	0.02497
10. Fifth Cosine harmonic	0.05789	0.02405
11. Interaction of contemporaneous temperature with annual sine harmonic	0.40596	0.18536
12. Interaction of contemporaneous temperature with annual cosine harmonic	-0.20666	0.23336
13. Deviation from ln(31 or 61 day moving average of maximum daily air temperature)	0.46657	0.13966
14. Interaction of contemporaneous rain with annual sine harmonic	-0.07925	0.01327
15. Interaction of contemporaneous rain with annual cosine harmonic	-0.03511	0.01598
16. Deviation from ln(31 or 61 day moving sum of rainfall+1)	-0.04401	0.01160
17. Bi-monthly lag from rain deviation	-0.03207	0.00701
18. Constant (Mean intercept)	5.48743	0.00385
Standard Error of Individual Constant Terms	0.63531	
Standard Error of White Noise Error	0.30522	
Number of observations	8580	
Number of customer accounts	305	
Time period	**January 1992- August 2000**	

REFERENCES

Altman, T., Heath, E., Boulos, P., Orr, S., Automated Calculation and Distribution of Water Demands in Network Models, Proc. AWWA Computer Conference, Austin, Texas, pp. 124-129, 1997.

American Water Works Association (AWWA), 1989. AWWA Manual M32. AWWA.

Billings, R. B. 1982. "Specification of Block Rate Price Variables in Demand Models," Land Economics, vol. 58, p. 386-394.

Billings, R. B. and C.V. Jones. 1996. Forecasting Urban Water Demand. AWWA. Denver.

Baumann, Boland, and Hanemann 1988. Urban Water Demand Management and Planning, McGraw-Hill, Inc.

Bowen, P. T., J.F. Harp, J.W. Baxter, and R.D. Shull. 1993. Residential Water Use Patterns. Awwa Research Foundation.

Brittain, R.L. 1974. Small Meter Periodic Test Limit Extension Study. Unpublished report, Philadelphia Suburban Water Company, Philadelphia, PA.

Cesario, L. 1995. Modeling, Analysis, and Design of Water Distribution Systems. AWWA.

Clark and Grayman, 1998, *Modeling Water Quality in Drinking Water Distribution Systems*, AWWA, Denver Co.

Diba Consulting Software Engineers. 2000. Water Demand Forecaster v. 1.3 presentation. http://www.dcse.com.

Ford, L. R. and D. R. Fulkerson, 1962, *Flows in Networks*, Princeton University Press.

Frey, McCuller, Henke, Spackman, and Murphy. 2001. "Optimizing a water master plan." *Journal AWWA*, Vol. 93, No. 5, p. 86.

Goldberger, A.S. 1968. "The Interpretation and Estimation of Cobb-Douglas Functions," *Econometrica*, vol. 35, p. 464-472.

Graeser, H.J. 1958. Detecting Lost Water at Dallas. *Journal. AWWA*, 50(7):925-932.

Hanemann, W. M. 1984. "Discrete-Continuous Models of Consumer Demand," Econometrica, 1984.

Hannan E J. 1960. Time Series Analysis., Methuen.

Hsiao. 1986. *Analysis of Panel Data,* Cambridge University Press.

Hudson, W.D. 1964. Reduction of unaccounted-for Water. *Journal. AWWA*, 56(2):143-148.

Jones C.V. and Morris 1984 "Instrumental Price Estimates and Residential Water Demand." Water Resources Research, vol. 20, p. 197-202.

Jorgenson 1964 "Minimum Variance, Linear, Unbiased Seasonal Adjustment of Economic Time Series," Journal of the American Statistical Association, Vol. 59, No. 307, September 1964, pp. 681-724.

Kuranz, A.P. 1942. Meter Maintenance Practice. *Journal. AWWA*, 34(1):117-120.

Kruse, W.E. 1974. *Determination of Water Consumption,* Internal Los Angeles Department of Water and Power Lecture.

Lowry, S.G. and Porter, R.S. 1994. Automated Demand Allocating, Updating, and Forecasting Through Integration of GIS, Distribution System Model and Erie County, New York, Water Authority Customer Database. In *Proc. Computers in the Water Industry Computer Conference.* AWWA.

Mayer, P.W., and DeOreo, W.B., 1997. North American Residential End Use Study: Progress Report. 1997 AWWA Annual Conference.

Mundlak, Y. 1978. "On the Pooling of Time Series and Cross Section Data," Econometrica, vol. 46, p. 69-85

Nielsen, N. 1969. Determination of Proper Age for 5/8 Inch Meters in the Hackensack and Spring Valley Systems. *Report for the Hackenstack Water Co.*, Harrington Park, NJ.

Newman, G.J., and Noss, R.R. 1982. Domestic 5/8 Inch Meter Accuracy and Testing, Repair, and Replacement Programs. In *Proc. Of the 1982 AWWA Annual Conference*, Part 1, Paper No. 11-7. Denver, CO: AWWA.

Orr, C.H., Alegre, H., and Coelho, S.T. 1994. System Analysis and Modeling for a Water Information System. In *Proc. Computers in the Water Industry Computer Conference,* AWWA.

Purves, A. and Lee Cesario. 1993. "Computer Applications in the Water Industry," Journal AWWA, Apr. 1993, p129-130.

Sisco, R.J. 1967. The Case of Meter Replacement Programs. *Journal. AWWA*, 59(11):1149-1455.

Wooldridge. 2002. *Econometric Analysis of Cross Section and Panel Data.* The MIT Press

Yanov, D.A., and Koch, R.N. 1987. A Modern Residential Flow Demand Study. Session 27 of AWWA Annual Conference, Kansas City, KS.

Zheng Y. W., P. F. Boulos, C. H. Orr, and J. J. Ro. 1998. "Using Genetic Algorithms to Rehabilitate Distribution Systems," *Journal AWWA*, Vol. 98, No. 11, p. 74.

ABBREVIATIONS AND ACRONYMS

AF	acre feet
AWWA	American Water Works Association
AwwaRF	Awwa Research Foundation
ccf	hundred cubic feet
CF	cubic feet
d	day
gal	gallon
GIS	geographic information system
gpd	gallons per day
in.	inch
LADWP	Los Angeles Department of Water and Power
mgd	million gallons per day

IWA Publishing's authorised EU representative for General Product Safety Regulations is Diane D'Arras, 15 rue Duret, 75116 Paris, France, e-mail: safety@iwap.co.uk.

Printed and bound by CPI Group (UK) Ltd, Croydon, CR0 4YY
17/09/2026
02218802-0010